Energy, Force, and Matter

CAMBRIDGE HISTORY OF SCIENCE

Editors

GEORGE BASALLA
University of Delaware

WILLIAM COLEMAN
The University of Wisconsin

Energy, Force, and Matter

The Conceptual Development of
Nineteenth-Century Physics

P. M. HARMAN

Department of History, University of Lancaster

CAMBRIDGE UNIVERSITY PRESS
CAMBRIDGE
LONDON   NEW YORK   NEW ROCHELLE
MELBOURNE   SYDNEY

Published by the Press Syndicate of the University of Cambridge
The Pitt Building, Trumpington Street, Cambridge CB2 1RP
32 East 57th Street, New York, NY 10022, USA
296 Beaconsfield Parade, Middle Park, Melbourne 3206, Australia

First published 1982

Printed in the United States of America

Library of Congress Cataloging in Publication Data
Harman, P. M. (Peter Michael), 1943–
Energy, force and matter.
(Cambridge history of science)
Bibliography: p.
Includes index.
1. Physics – History. 2. Force and energy – History.
I. Title. II. Series.
QC7.H44   530'.09'034   81–17029
ISBN 0 521 24600 8 hard covers   AACR2
ISBN 0 521 28812 6 paperback

CHAPTER I

Introduction:
The Conceptual Structure
of Nineteenth-Century Physics

In the nineteenth century the term 'physics' acquired new and significant connotations. Although the term was still occasionally used in the traditional sense to refer to natural science in general, by the early nineteenth century 'physics' was being used in the modern and more specialised sense to denote the study of mechanics, electricity, and optics, employing a mathematical and experimental methodology. In the article entitled 'Physical Sciences' in the ninth edition of the *Encyclopaedia Britannica* in the 1870s, James Clerk Maxwell identified the scope of physics with the programme of mechanical explanation, first enunciated in the seventeenth-century 'mechanisation of the world picture', which sought to explain physical phenomena in terms of the structure and laws of motion of a mechanical system. In a critical exposition of current physical theory, *The concepts and theories of modern physics* (1881), Johann Bernhard Stallo gave an informative and more detailed definition of the theoretical structure of physics as conceived by contemporary theorists:

The science of physics, in addition to the general laws of dynamics and their application to the interaction of solid, liquid and gaseous bodies, embraces the theory of those agents which were formerly designated as imponderables – light, heat, electricity and magnetism, etc.; and all these are now treated as forms of motion, as different manifestations of the same fundamental energy.

In the nineteenth century the science of physics came to be defined in terms of the unifying role of the concept of energy and the programme of mechanical explanation.

The concept of energy provided the science of physics with a new and unifying framework and brought the phenomena of

1

physics within the mechanical view of nature, embracing heat, light, and electricity, together with mechanics, in a single conceptual structure. The major theme of the development of physics in the nineteenth century is the way in which theoretical innovations – the concept of the physical field, the theory of the luminiferous and electromagnetic ether, and the concepts of the conservation and dissipation of energy – were formulated according to the mechanical view of nature, which supposed that matter in motion was the basis of all physical phenomena.

This book will focus on key themes that defined the structure of physics in the nineteenth century. I will begin with an account of the development, by around 1850, of a distinctive science of physics that took quantification and the search for mathematical laws as its universal aims and that established the law of the conservation of energy as a unifying principle and mechanical explanation as the programme of physical theory. The structure of physical theory in the nineteenth century will then be analysed by focusing on the status of the concepts of energy, force, and matter in the physics of the period. I will describe the development of the principles of the conservation and dissipation of energy, the theory of the physical 'field' (which accounted for the transmission of force by means of the mediating action of the 'field' between bodies), and the study of molecular physics. A preliminary outline of the scope of the argument will provide an introductory survey of the conceptual development of physics in the nineteenth century.

The Context of Physical Theory

In eighteenth-century physical theory mechanical phenomena were studied mathematically, and hypotheses about atoms and the nature of forces were avoided; by contrast, heat and electricity were generally explained by supposing imponderable 'fluids' of heat and electricity and forces acting between the particles of these 'fluids' and the atoms of ordinary matter. These speculative and generally qualitative theories stood apart from the exact, quantitative science of mechanics, though by the late eighteenth century attempts were made to treat heat and electricity mathematically, attempts that initiated the conceptual unification of the science of physics. The creation of a unified physics was fostered by four significant developments.

1. P. S. de Laplace and his followers formulated a mathematical theory of interparticulate forces, to be applied to mechanical as

well as to thermal and optical phenomena. Although this theory was displaced by new developments in heat and optics in the decade 1815–25, the Laplacian emphasis on mathematisation and the formulation of a unified physical world view had an important impact on the subsequent development of physical theory.

2. The publication of Joseph Fourier's mathematical theory of heat in 1822 brought the study of heat within the framework of mathematical analysis previously applied only to mechanical problems. In bridging this traditional conceptual dichotomy, and in stressing the distinction between mathematical and physical representation, Fourier's work had profound general implications for the creation of a unified physics. In the 1840s, influenced by the mathematical analogy between Fourier's theory of heat and the theory of electrostatics, William Thomson explored the mathematical and physical analogies between, on the one hand, the laws of heat and electricity and, on the other, the mechanics of particles and fluid and elastic media. Thomson's use of the method of physical analogy, in which the common mathematical form highlighted the conceptual relations between disparate phenomena, emphasised the unity of the phenomena of physics.

3. A. J. Fresnel's wave theory of light, which supposed that light was propagated by the vibrations of a mechanical ether, brought optics within the framework of the mechanical view of nature. By the 1830s the wave theory of light was generally accepted, and physicists explored a variety of physical and mathematical theories in an attempt to provide a coherent mechanical theory of optics. The mechanical theory of the optical ether established a paradigm for the programme of mechanical explanation.

4. The formulation of the law of the conservation of energy in the 1840s stressed the unity of physics, subsuming the phenomena of heat, light, electricity, and magnetism within the framework of mechanical principles. In the early nineteenth century, physicists explored the interconversion of light and heat, and of electricity and magnetism; and the experiments of H. C. Oersted in 1820 and Michael Faraday in 1831, which established the connections between electric and magnetic forces, were of especial importance in justifying the doctrine of the unity and convertibility of natural 'powers' or 'forces', an idea that was reformulated as the principle of the conservation of 'energy' in the 1840s. James Prescott Joule's experiments established the equivalence of heat and mechanical work; and in a seminal essay of 1847, Hermann von Helmholtz expressed the relation among mechanics, heat, light, electricity,

and magnetism by treating these phenomena as different manifestations of energy. Helmholtz formulated the law of the conservation of energy as a mathematical and mechanical theorem, emphasising the unifying role of the energy concept as an expression of the mechanical view of nature.

By 1850 the law of the conservation of energy had provided a new framework for physical theory based on the mechanical view of nature, which rejected the supposition of anomalous forms of matter and supposed that particles of ordinary matter in motion should be considered the basis for physical theory. The physical problems of light, heat, and electricity were conceptualised in a way that made them amenable to mathematical analysis and thereby fostered the unification of physics.

Energy Physics and Mechanical Explanation

The study of the relationships between heat and mechanical work was of central importance in nineteenth-century physics. The formulation of the laws of 'thermodynamics' bridged the disjunction between mechanics and heat and helped to establish the dominance of the mechanical view of nature. Whereas eighteenth-century physicists had considered mechanical and nonmechanical processes as separate physical systems, Joule's demonstration of the equivalence of heat and mechanical work in the 1840s, and the formulation of the law of the conservation of energy, established the unification of mechanical and thermal processes.

The science of thermodynamics was concerned with the direction of heat flow in the production of work, as well as the establishment of the principle of the equivalence of heat and work. In formulating the conceptual basis of thermodynamics in 1850, Rudolf Clausius resolved a problem that had been raised by Thomson, an apparent conflict between Joule's claim that heat was consumed in the generation of mechanical work and the theory of heat engines that had been proposed by Sadi Carnot in 1824. Carnot had argued that the crucial factor in the generation of work by a heat engine was the temperature difference in the engine: Work was generated by the passage of heat from a warm to a colder body, heat being conserved in the process. Clausius established that Carnot's theory that heat passes from a warm body to a colder one whenever work is done by a heat engine was consistent with Joule's assertion that whenever work is produced by heat, a quantity of heat proportional to the work generated is consumed, if Carnot's assumption that heat was conserved in the generation of

mechanical work by heat engines was abandoned. Joule's principle and Clausius's modified form of Carnot's theory provided the basis for the two laws of thermodynamics.

Thomson and Clausius formulated the science of thermodynamics in terms of the mechanical view of nature, and they maintained that the principle of the equivalence of heat and work was consistent with the mechanical theory of heat: that heat consists of the motions of the particles of bodies. While affirming his support for the mechanical view of nature, Thomson specifically avoided suggesting any mechanical model of thermal processes. Clausius, by contrast, ultimately sought to render the laws of thermodynamics intelligible by appealing to a theory of molecular motions.

The postulation of a model of molecular arrangement became fundamental to Clausius's interpretation of the second law of thermodynamics. In proposing his own account of thermodynamics in 1851, Thomson argued that the key issue was the explanation of irreversible thermal processes. In Thomson's view the second law of thermodynamics asserted the dissipation of energy in irreversible processes. The two laws of thermodynamics were consistent, for although energy was dissipated in irreversible processes, it was not destroyed but was merely transformed into other forms of energy. In the 1850s and 1860s Clausius sought to formulate concepts that would provide a measure of the direction of thermodynamic processes and clarify the nature of irreversible processes. His concept of 'entropy' denoted the directional character of physical processes: The law of the increase in entropy became the familiar form in which the second law of thermodynamics was expressed. Clausius sought to explain entropy by grounding it on a mechanical model of molecular arrangement and motion, a model that for him had a more fundamental status than the entropy concept itself.

Although the relationship between entropy and models of molecular arrangement became the subject of debate, physicists were agreed that the laws of thermodynamics provided an expression of the mechanical view of nature. Thomson maintained that all forms of energy were forms of mechanical energy, and he strove to establish the energy principle as the kernel of the mechanical view of nature. In the 1850s and 1860s Thomson and W. J. Macquorn Rankine elaborated a framework of physical theory based on the primacy of the energy concept, attempting to clarify the mathematical and physical basis of the principle of the conservation of energy as a reformulation and generalisation of the doctrine of the convertibility of natural 'forces'. Thomson and Peter Guthrie Tait's

Treatise on natural philosophy (1867), based on energy and mechanical explanation, became the text of the new framework of physics, establishing energy within the conceptual structure of the theory of mechanics.

Ether and Field Theories

The concept of the physical 'field', which supposed that electric and magnetic forces were distributed in space and mediated by the agency of a physical field, had become established in British physics by around 1850; its further development was bound up with the elaboration of mechanical theories of the ether to explain the physical constitution of the field. In the 1830s and 1840s Faraday had argued that electric forces were transmitted between particles in an ambient medium, and used the concept of 'lines of force' to represent the disposition of electric and magnetic forces in space. These theories provided the conceptual basis for the notion of the mediating agency of the field existing in the intervening spaces between electrified and magnetic bodies. In formulating theories of electricity and magnetism in the 1850s, Thomson and Maxwell appealed to the field concept rather than to the supposition of 'distance' forces acting directly between electrified and magnetic bodies across finite distances of space.

To explain the physical structure of the field, Thomson proposed that its action could be represented by molecular vortices in the ether; later, however, he conceived the ether as a plenum, representing the field of force by an ethereal continuum. In the 1850s and 1860s Maxwell formulated a series of physical and mathematical theories of the field, drawing on Faraday's physical concepts as well as Thomson's notion of molecular vortices in elaborating a mechanical model that represented the action of the field in transmitting forces by the action of particles in the ether. Maxwell's physical ether theory of 1861–2 provided a systematic theory of the propagation of electric and magnetic forces, employing a mechanical ether that was intended as an illustrative model rather than an ultimate physical explanation. He refined his theory of the field in a paper published in 1865, and though he continued to uphold his mechanical interpretation, emphasising that electromagnetic phenomena were produced by the motions of particles of matter in the ether, he abandoned any attempt to formulate a specific mechanical model of the field, employing instead the methods of Lagrangian analytical dynamics, a generalised formalism not linked to any specific mechanical model.

Maxwell's theory of the electromagnetic field had an unexpected implication: that the velocity of electromagnetic waves propagated in the ether was identical to the velocity of light. This led Maxwell to the identification of the electromagnetic and luminiferous ethers and to his 'electromagnetic theory of light', the concept of light as an electromagnetic vibration in the ether, a unification of optics and electromagnetism that was grounded on a mechanical theory of the ether. The experimental detection of electromagnetic waves by Heinrich Hertz, announced in 1888, was immediately viewed as a striking confirmation of the electromagnetic field. To Continental physicists, especially, it served to establish Maxwell's field theory of electromagnetism in place of the various versions of the action-at-a-distance theory that had been developed by German physicists in the nineteenth century.

By the 1890s the concept of the physical field was subject to a variety of interpretations. Some British physicists tried to integrate Maxwell's theory of the electromagnetic field and Thomson's concept of the ethereal continuum; several physicists developed formulations of the electromagnetic field equations that were not based on a mechanical ether model. The most radical approach was adopted by H. A. Lorentz, who proposed a universal physics grounded on purely electromagnetic concepts. Lorentz's theory also provided an explanation of the experiments performed by A. A. Michelson and E. W. Morley in the 1880s, which had raised serious difficulties in the explanation of the relationship between ether and matter. In Lorentz's 'electron' theory, matter was conceived in terms of charged particles (electrons), and the relationship between ether and matter was explained as the connection between electrons and the electromagnetic field. Lorentz envisaged an electromagnetic rather than a mechanical view of nature, denuding the ether of mechanical properties. By 1900 developments in ether and field theory challenged the hegemony of the mechanical view of nature.

Problems of Molecular Physics

The mechanical view of nature received additional support in the 1850s and 1860s, with the development by Clausius and Maxwell of the kinetic theory of gases, which supposed that gases consisted of particles of matter in motion. Physicists began to stress the molecular theory of matter. Throughout the latter part of the nineteenth century they based speculations about the properties of matter on the kinetic theory of gases, but it became apparent that

this theory could not satisfactorily explain molecular structure. Moreover, the evidence of spectroscopy seemed to conflict with conclusions about molecular structure derived from the theory of gases. These problems led to much debate about the nature of molecular models, and doubts concerning the status of the kinetic theory challenged the coherence of the mechanical view of nature.

The problems of the molecular theory of matter also bore on the interpretation of thermodynamics. Maxwell's theory of gases was based on his introduction of a statistical theory of molecular motions, and he invoked this theory in discussing the second law of thermodynamics. Maxwell introduced his 'demon' paradox to show that any molecular interpretation of the second law must be based on a statistical analysis of the motions of an immense number of molecules. There were continual spontaneous fluctuations of individual molecules, he noted, in which heat was transferred from a cold body to a hotter one by the random motions of molecules, but these fluctuations did not constitute a violation of the second law of thermodynamics. This law, an essentially statistical one, applied to an immense number of molecules, not to the behaviour of individual molecules; and hence any interpretation of it that was based on a theory of the motions of individual molecules (as suggested by Clausius) was misconceived.

In a seminal paper of 1877, Ludwig Boltzmann established the relationship between entropy and the statistical analysis of molecular motions, characterising the irreversible increase of entropy in natural processes as a statistical law. Boltzmann sought to defend the mechanical view that nature consisted of particles of matter in motion by explaining the second law of thermodynamics in terms of a statistical theory of molecular motions. In the 1890s this interpretation of thermodynamics came under attack. Max Planck emphasised the absolute validity of the entropy concept, and criticised Boltzmann's interpretation of entropy as a statistical concept. Planck questioned the intelligibility of an explanation of entropy drawn from the kinetic theory of gases. This denial of the idea that entropy should be explained in terms of mechanical principles of matter in motion challenged the whole programme of mechanical explanation.

The Status of Mechanical Explanation

Though the programme of mechanical explanation came under attack in the 1890s, the mechanical or, as it was often termed, the 'dynamical' world view, which supposed an ontology of particles of

matter in motion as the substratum underlying physical reality, dominated physical theorising in the nineteenth century. I will use the term 'ontology' to denote assumptions about the basic constituents of physical reality, as distinct from specific hypotheses or models about reality. The ontology of the 'dynamical' theory of particles of matter in motion was fundamental to the dominant programme of physics in the nineteenth century, the explanation of physical phenomena by the structure and laws of motion of a mechanical system. Nevertheless, physicists recognised the gap between the presupposition of the ontology of the mechanical world view and the invention of hypothetical mechanical models to represent physical phenomena; and the relationship between physical reality and the symbolic representations employed for its depiction was a theme of fundamental importance. Many physicists stressed the gap between the structure of physical reality and the encompassing net of theory, and discussions about the conceptual status of the mechanical models employed to represent phenomena were interwoven with debates on the nature of physical reality, and shaped the interpretation of mechanical explanation as a coherent programme.

Physicists used mechanical explanation in three ways. The first appealed to theories of the configuration and motion of particles of matter, aiming to explain natural phenomena by the arrangement of particles of matter and the forces acting between particles. The second sense of mechanical explanation involved the postulation of mechanical models, either the depiction of hypothetical models involving wheels and springs or the construction of working mechanical devices as representations of phenomena. These mechanical models were not necessarily envisaged as representations of reality, but were seen as demonstrating that phenomena could in principle be represented by mechanisms; mechanical construction rendered phenomena intelligible. The third sense in which mechanical explanations were formulated involved an attempt to avoid speculation about the physical structure of the mechanical system supposed to represent the phenomena. Theorists using this approach held that it was impossible to elaborate a unique mechanical model of any phenomenon, and appealed to the abstract formalism of Lagrangian analytical dynamics. The equations of motion obtained in this way were independent of the structure of the connections of the mechanical system, but phenomena were nevertheless subsumed under the principles of mechanical explanation, though not represented by a specific, visualisable mechanical model.

The tension between physical and mathematical models of

mechanical systems, and the relationship between mechanical representations and physical reality, played an important role in shaping nineteenth-century theories of physical reality. Maxwell in particular discussed these problems in a sophisticated and influential manner, and debates about the appropriate assumptions to guide the formulation of physical theories underlay the elaboration of new theories in the period. Aware of the gap between theory and reality, many nineteenth-century physicists noted the limitations as well as the objectives of the programme of mechanical explanation.

The Historiography of Physics

The profound conceptual changes in physics in the twentieth century, the abandonment of the doctrines of absolute space and time in Einstein's theory of relativity and of causality and determinism in quantum mechanics, have customarily led to a depiction of the development of physics by means of a disjunction between 'classical' or 'Newtonian' and 'modern' physics. The interpretation of eighteenth- and nineteenth-century physics as monolithic and possessing a unified conceptual structure, its domination being curtailed only by the advent of 'modern' physics, is deeply embedded in the historiography of science. This traditional interpretation is sustained by an account of the 'Scientific Revolution' of the seventeenth century that misrepresents the implications of this intellectual revolution for later scientific developments. The Scientific Revolution has been traditionally characterised as a philosophical revolution in which explanations of natural phenomena came to be couched in mechanical laws. The 'mechanisation of the world picture' is depicted as culminating in the Newtonian synthesis of mechanics and astronomy, and the establishment of Newtonian mechanics and of the programme of mechanical explanation is seen as providing the framework for 'classical' physics in the following two centuries.

Although the shifts in attitude in seventeenth-century conceptions of nature were profound, it is misleading to label this transition the Scientific Revolution if this label is meant to carry the implication that later developments can be comprehended in terms of the scientific categories propounded in this period. The term 'Newtonian' as applied to eighteenth- and nineteenth-century physics implicitly conflates Newton's natural philosophy and the physics of this later period, and is hence a misleading designation. The developments in theoretical mechanics in the eighteenth

Contents

Preface

The period circa 1800–1900 corresponds to a distinctive phase in the conceptual development of physics, bounded by the increasing dominance, from the late eighteenth century on, of quantification and the search for mathematical laws, together with the emergence of a unified physics based on the programme of mechanical explanation, and by the development in the early twentieth century of the quantum and relativity theories. I have aimed to provide a study of the development of physics in the nineteenth century in a form accessible to the reader without a specialised knowledge of physics and mathematics. The argument of the book is structured around the major conceptual problems of nineteenth-century physics: the emergence of energy physics and thermodynamics, the theory of the luminiferous and electromagnetic ether and the concept of the physical field, molecular physics and statistical thermodynamics, and the dominance of the programme of mechanical explanation. The book begins with an account of the transformation in the scope of the science of physics in the first half of the nineteenth century.

I am grateful to John Heilbron for reading a portion of the manuscript and to Crosbie Smith for reading the whole manuscript of this book, and for their helpful comments. I am also grateful to the Syndics of the Cambridge University Library for their kind permission to reproduce documents in their keeping, and to the Council of the Royal Society for the award of a grant for research undertaken in the preparation of this book.

For Juliet and Timothy

century show a significant departure from the mechanical and mathematical assumptions of Newton's natural philosophy; and the physics of imponderable 'fluids', active substances, and anomalous forms of matter current in the eighteenth century contrasts with Newton's theory of nature, though Newton's speculative writings influenced the shape of these physical theories. Despite the dominance of the programme of mechanical explanation, and the occasional appeal to a 'Newtonian' theory (as in the Laplacian theory of attractive and repulsive forces), the term 'Newtonian' is misleading when applied to physics in the nineteenth century. The conceptual innovations of nineteenth-century physics – energy conservation, the theory of the physical field, the theory of light as the vibrations of an electromagnetic ether, and the concept of entropy – cannot meaningfully be described as 'Newtonian'. The image of classical physics as monolithic, forming a unified world view, ignores the striking developments in eighteenth- and nineteenth-century physics. Although 'classical' is appropriate when used to distinguish the philosophical assumptions of eighteenth- and nineteenth-century physics from the relativistic and indeter-ministic doctrines of twentieth-century physics, the term 'New-tonian' is misleading as a characterisation of the structure and content of nineteenth-century physics; its use blurs the distinctive conceptual principles and physical world view of the period.

CHAPTER II

The Context of Physical Theory: Energy, Force, and Matter

In style and content, the physical theory of 1850 shows a marked contrast to that prevalent in 1800. By 1850 the limits and internal cohesion of the science of 'physics' were clearly articulated, and the subject had achieved a new and well-defined conceptual content and unity. By 1850 some of the main themes of nineteenth-century physics had been formulated: the unification of physical phenomena within a single explanatory framework, the primacy of mechanical explanation as an explanatory programme, the mathematisation of physical phenomena and the role of mathematical analogy as a guide to the formulation of physical theories, and the enunciation of the principle of energy conservation as a universal, unifying law. The emergence of these broad and unifying themes contrasts with the disunity in physical theory in 1800.

The general disjunction in eighteenth-century physical theory can be illustrated by a contrast between Newton's *Philosophiae naturalis principia mathematica* [*Mathematical principles of natural philosophy*] (1687) and his *Opticks* (1704). In the *Principia* Newton offered the paradigm of a mathematical science of 'rational mechanics', and though he expressed the hope that all physical phenomena could be subsumed under analogous mathematical methods (illustrating his intentions by a mathematical treatment of optical refraction), in the *Opticks* he based his treatment of the problems of optics and chemistry on an experimental methodology and a speculative theoretical structure, an atomistic physics that became bloated in later editions to include a variety of explanatory agents, forces, active principles, and the ether. This disparity of methods and models was echoed in eighteenth-century writings on physical theory.

Newton's mathematical theory of nature in the *Principia* was grounded on the revolution in mathematics of the sixteenth and seventeenth centuries. The analytic geometry that Newton imbibed from Descartes had led to a shift from the visual to the abstract in mathematics; visual representation was replaced by equations expressing relations between geometrical quantities. The calculus methods developed independently by Newton and Leibniz, concerned with the treatment of infinitesimal quantities, proved amenable to the solution of the physical problems of mechanics, especially the study of forces and of changes of motion. In his *Principia* Newton represented the curve traversed by the motion of a planet under the influence of a gravitational force as a sequence of infinitesimal line segments, the attractive force being considered as a series of discrete force impulses. The work of Continental mathematicians, using the Leibnizian differential calculus, proved brilliantly successful in the solution of mechanical problems. Relations between physical quantities were expressed as relations between geometrical quantities, represented by algebraic symbols. The separation of analysis from geometry in the eighteenth century led to the development of flexible methods for the mathematical expression of physical quantities. Mathematical symbols were stripped of their geometric foundations and employed directly to represent physical quantities, a procedure that enabled complex physical concepts to be represented mathematically. The Continental geometers, including Johann Bernoulli, his son Daniel Bernoulli, Jean d'Alembert, and Leonhard Euler, performed sophisticated mathematical analyses of mechanical problems of fluid and elastic media. In the tradition of 'rational mechanics', the mathematical laws of mechanics were formulated on the basis of experiential concepts of mass, length, and time; and unobservable, hypothetical explanatory entities were avoided.

The mathematisation of physical phenomena achieved its major successes in the study of mechanics. Although both the Newtonian 'emission' theory of light as particles subjected to forces acting between the particles of ordinary matter and the light corpuscles, and Euler's theory of light as the propagation of a pulse in a fluid, were based on mathematical arguments, light was frequently regarded as analogous to 'fire' and represented as an imponderable fluid 'ether'. These 'fluids' were envisaged as composed of mutually repelling particles (this property being denoted as 'elasticity'), as being 'subtle' (being able to penetrate the empty spaces between the particles of ordinary matter), and as being attracted by ordinary matter. This theory was applied to electricity, heat, and chemistry, which were explained by the suppositions of interparticulate forces and ethereal fluids.

The Newtonian ether and interparticulate forces, especially as proposed in the conjectural 'Queries' appended to the various editions of the *Opticks,* played an important role in shaping the development of these theories. The early Newtonian commentators emphasised Newton's concept of short-range attractive forces, a model that provided the basis for a theory of chemistry that envisaged a quantified physics of chemical forces, the diverse chemical elements being considered as different structures of ultimate atoms linked by forces. By the middle of the eighteenth century this theory was firmly established in the writings of French chemists seeking a science of chemical forces ('affinities') as the key to the explanation of chemical reactions; and though there were attempts to quantify this theory of chemical forces, it remained largely qualitative. By the 1740s attention was being paid to Newton's concept of repulsive forces, under the impetus of renewed interest in Newton's 'subtle', 'elastic' ether, the particles of which were endowed with repulsive forces. The Dutch chemist Herman Boerhaave's concept of 'fire' as an active principle counteracting the attractive forces of ordinary matter also excited interest and led to the elaboration of dualistic conceptions of nature. Benjamin Franklin's theory of electricity posited a dualism of ordinary matter endowed with attractive forces and a repellent electrical 'atmosphere' surrounding the particles of ordinary matter; electrification was represented by the permeation of the electrical 'fluid' through the interstitial pores of the electrified body. This electrical matter was analogous both to Newton's ether and to Boerhaave's fire. By the end of the century, imponderable fluids were employed to explain electricity, magnetism, light, and heat. Although many theorists invoked a multiplicity of fluids to explain these different phenomena – one or two fluids for electricity, one for heat, 'phlogiston' as the imponderable principle of combustion – others proposed a unified ether theory, in which the imponderable fluids were regarded as various modifications of the ether; in this dualistic world view the transformations of an ethereal, active, repellent substance balanced the attractive power of ordinary matter.

In the eighteenth century the study of the phenomena of heat, magnetism, and electricity remained largely qualitative, but towards the end of the century there were attempts to subject these phenomena to quantitative and mathematical treatment, a development that was fostered by improvements in the precision of scientific instruments and the increasing professionalisation of physics. The work of Joseph Black and of A. L. Lavoisier and Laplace on heat, of Tobias Mayer, J. H. Lambert, and C. A. Coulomb on magnetism, of Alessandro Volta, Henry Cavendish, and Coulomb on electrostatics, all employed precise measurement and quantification as criteria of theory construction. The quantification of electrostatics, which established the law of electrostatic force, was especially important in providing the paradigm of precise experimentation, quantification, and the search for mathematical laws as the methodology and objectives that characterised nineteenth-century physics.

Laplacian Physics

The theory of interparticulate forces and imponderable fluids attained its most comprehensive form in the work of Pierre Simon de Laplace (1749–1827) and his school in the first two decades of the nineteenth century. Laplace argued that optical refraction, the cohesion of solids, capillary action, and chemical reactions were all the result of an attractive force exerted by the particles of matter, and maintained that the enunciation of the short-range interparticulate force law would bring the study of terrestrial physics to the level of perfection that Newton's law of universal gravitation had attained for the study of celestial physics. These objectives were not original: A. C. Clairaut had discussed a mathematical theory of short-range molecular forces in his work on optical refraction and capillary action in the 1730s and 1740s. Nevertheless, Laplace's treatment of refraction and capillary action in the 1805 volume of his *Traité de mécanique céleste* [*Treatise on celestial mechanics*] and its supplements published in the following two years provided a systematic and mathematical account of these phenomena in terms of molecular forces. Laplace stressed the universality of this explanatory programme, later observing that quite apart from its application to thermal and optical phenomena and the theory of capillary action, 'it would be useful to introduce such a study [of molecular forces] in proofs in mechanics'. By implication, Laplace rejected the tradition of rational mechanics in favour of a new universal physics based on the hypothesis of molecular motions and forces, a theory of nature applicable to the problems of mechanics as well as to the optical, thermal, and electrical phenomena that had traditionally been explained by means of an appeal to interparticulate forces. The principles of this programme were stated most

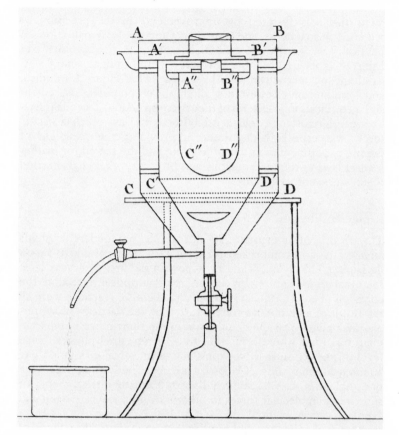

Fig. 2.1. The 'ice-calorimeter' of Lavoisier and Laplace (1783). The apparatus was about three feet high and consisted of three cylindrical metal containers nesting in one another. The outermost vessel held crushed ice, to insulate the inner vessels from external heat. A wire basket holding the experimental substance was suspended in the inner container, in contact with an inner layer of ice that was melted by the heat-producing substance in the basket. The water produced by melting the ice was drained off and weighed, the quantity of ice melted being proportional to the heat emitted by the substance. Lavoisier and Laplace determined the specific heats of a variety of substances, that is, the quantities of heat required to raise equal masses of different substances by the same number of degrees, relative to the specific heat of water as a standard of reference and taken as unity. The method, if not the apparatus, was the work of Laplace.

The experiment provided an important example of precise experimentation to achieve numerical measures of physical quantities, an ideal of

explicitly by Siméon Denis Poisson (1781–1840) as a theory of 'physical mechanics' that would replace the 'analytical mechanics' cultivated by eighteenth-century mathematicians in the study of mechanical problems. Poisson stressed that the theory of molecular forces would be applied to such mechanical problems as the study of flexible strings, elastic surfaces, and the pressure of fluids.

The Laplacian school's emphasis on unification, bridging the disjunction between mechanics and the phenomena of heat, light, and electricity, had an enduring impact on the development of physics in the nineteenth century, even though the Laplacian concept of physical mechanics based on the supposition of molecular forces fell out of favour. Laplace's encouragement of precise experimental work as a counterpart to his mathematical physical theory was of equal importance. In his influential *Traité de physique expérimentale et mathématique* [*Treatise on experimental and mathematical physics*] (1816), Jean Baptiste Biot (1774–1862) affirmed the importance of precise experimentation and the establishment of numerical measures of physical quantities, stressing the need to improve the accuracy and precision of experiments by introducing new procedures and instruments. Biot presented a mathematical and experimental method as the paradigm for physics; quantification was to be the aim of physical theory.

Laplace envisaged a quantified physics that would embrace the theory of imponderable fluids. His theory of heat and gases was based on the supposition of an imponderable fluid of heat surrounding the particles of matter, a fluid that had been termed 'caloric' by Lavoisier and his colleagues. The properties of caloric were supposed to be modified by its state of combination with ordinary matter, the elastic properties of gases being attributed to their possession of more caloric between their particles than was the case for solids and liquids. Laplace considered that the relation-

---

Fig. 2.1. (*cont.*)
quantification stressed by Biot in his comprehensive survey of physics. In their paper, Lavoisier and Laplace emphasised that they did not hold any hypothesis about the nature of heat, but they later became committed to the 'caloric' theory of heat, which was favoured by Biot. The interest in calorimetry, and the development of concepts of the quantity of heat and of the specific and latent heats of substances, encouraged the establishment of the concept of heat as caloric, associated with or entering into combination with material substances. *Source*: Jean Baptiste Biot, *Traité de physique expérimentale et mathématique*, 4 vols. (Paris, 1816), 4:pl. V, fig. 66.

ship between the attractive forces exerted by the particles of ordinary matter and the repulsive forces exerted by the particles of caloric determined the properties of gases. The caloric theory was linked to Laplace's quantified physics, and achieved a sophisticated form in his theory of gases.

In urging the universal applicability of his programme of molecular physics, Laplace drew upon the work of his associate Claude Louis Berthollet (1748–1822). Berthollet argued that chemical affinity was the result of attractive forces between the particles of matter, declaring it probable that chemical affinity and gravitational attraction were the same property. Berthollet's detailed study exposed the inherent difficulties in constructing tables of chemical affinities to provide both a classification of the relative tendencies of chemical substances to combine with one another and a quantitative measure of the different attractive forces of chemical substances. He emphasised that the chemical activity of substances depended not simply upon their affinity but also upon their mass. The complexity of chemical phenomena jeopardised the programme of constructing a quantified physics of affinities, making it seem inapplicable. Nevertheless, Berthollet's systematic, mathematical chemistry provided a coherent theory of chemical affinities, a theory of molecular forces that probably helped to shape Laplace's statement of his programme of molecular physics.

Laplace exerted a dominating influence on the physics community in Napoleonic France, using patronage to direct research in support of his theory of molecular forces and encouraging quantitative experimental and mathematical studies of problems crucial to the success of his theory. Biot and François Arago (1786–1853) undertook an experimental study of optical refraction in gases, maintaining that their observations provided a measure of molecular forces. Etienne Louis Malus (1775–1812) presented an explanation on Laplacian lines of the phenomenon of double refraction, a solution to the problem that Laplace himself appropriated in urging the applicability of his theory of molecular forces to optical phenomena. In reporting his discovery of the polarisation of light, Malus likewise urged an explanation based on the hypothesis that molecular forces between the particles of light and ordinary matter produced the asymmetric properties of light.

Together with Berthollet's systematic theory of chemical affinities, Laplacian physical mechanics offered a unified and coherent programme of research, applying mathematically defined molecular forces to mechanics, heat, light, and chemistry. Despite its success and promise, this theory was soon attacked, and the challenge to it was fostered by the waning influence and patronage of Laplace and Berthollet with the fall of Napoleon in 1815. The theory of affinities was called in question by John Dalton's chemical atomism, which posited a quantification of the relative weights of the atoms of the chemical elements, rather than a mathematical theory of chemical forces. The caloric theory of heat came under attack, as did imponderable fluid theories in general; and the wave theory of light was proposed as an alternative to the corpuscular optics that had been central to the Laplacian theory of physics. Nevertheless, Laplacian physics had an enduring impact. In emphasising quantitative, exact experimental methods, and a mathematised unified physics that bridged the disjunction between mechanics and other physical phenomena, Laplacian physics established objectives, though not the structure of physical concepts, that were to dominate the creation of a unified science of physics in the nineteenth century. The ideals of mathematisation, quantitative experimentation, and a unified physical world view, together with the postulation of models of physical reality amenable to experimental test and mathematical formulation, shaped the development of nineteenth-century physics.

The Rejection of the Imponderable Fluids in British Physics

The abandonment of the imponderable fluids constituted one of the most significant developments in the transformation of physics in the early nineteenth century. In Britain the critique of imponderable fluid theories occurred in the context of the prevailing dualistic theory in which an imponderable, repellent ether balanced the attractive power of ordinary matter. In Britain the three notable critics of imponderable fluid theories were associated with the Royal Institution, founded in 1799 in response to an interest in the practical application of scientific research as an instrument of social and scientific improvement. Benjamin Thompson, Count Rumford (1753–1814), had rejected the theory of heat as an imponderable fluid in 1798, arguing that it could not explain the generation of heat by friction. The invention of the electric battery and the discovery of the electrical decomposition of water in 1800 led Humphry Davy (1778–1829) to enunciate a theory of electricity based on the forces of chemical affinity, abandoning the concept of the electrical fluid. The advocacy of the undulatory theory of light by Thomas Young (1773–1829), in a series of papers written between 1799 and 1804, led him to reject the concept of light as an elastic fluid analogous to fire.

Rumford rejected the imponderable fluid theory in favour of a theory in which the effects of heat were held to be the result of the interaction between the motions of the particles of ordinary matter and the vibrations of an ambient ether. Supposing a dualism of ordinary matter and the ether, he asserted that the vibrations of the ethereal 'atmosphere' surrounding the particles of ordinary matter were communicated to the surrounding ether and ultimately to other particles of ordinary matter. Rumford's dualistic theory recalls traditional ether theories, but he did not conceive the ether as an imponderable fluid; he emphasised that the vibrations of the ether, not the flow of an elastic fluid, gave rise to the effects of heat.

Young had adopted the theory of ethereal atmospheres surrounding the particles of ordinary matter in his early discussions of the mode of action of the ether, supposing a universal ethereal substance that he regarded as possibly a modification of the electric fluid, and arguing that light, heat, and the cohesive and repelling powers of bodies were due to the ether. Although he abandoned the attempt to construct a unified world view on the basis of an electrical ether, possibly because he could not satisfactorily explain cohesion and repulsion in these terms, the concept of a 'luminiferous' ether as the vehicle for the propagation of light and radiant heat became the central feature of his optics. Young's advocacy of the luminiferous ether and the undulatory theory of light led him to reject the Newtonian 'emission' theory of light as the projection of rays of discrete light corpuscles, as well as the concept of light as an ethereal elastic fluid analogous to fire. Young's principle of the interference of light waves, a concept developed from his earlier acoustical researches, used coalescing light waves to explain the diffraction of light, on the supposition that light waves could either reinforce or cancel each other to produce bands of lightness and darkness. Young transformed the theory of the unified ether into his wave theory of light.

In an early essay Davy had also signalled his commitment to the unified ether theory; and although he abandoned the concept of the electric fluid, the doctrine of the unity of natural powers characteristic of the unified ether theory continued to shape his physical theorising. He elaborated an electrical theory of chemical affinity, maintaining that electrical and chemical powers were so interconnected that it was likely that electrical and chemical forces were identical.

The theory of imponderable fluids was called into question increasingly in the early nineteenth century. The concepts of the balance of natural powers and the unity and interconversion of natural phenomena had a continued and significant influence on the development of physics in the nineteenth century, although these ideas were ultimately divorced from the imponderable fluids theory. Faraday and Joule developed the concept of the balance of powers into the theory of the convertibility and indestructibility of natural powers or 'forces', one of the conceptual strands that, transformed, became explicated by around 1850 as the conservation of energy.

Fresnel and the Elastic Solid Ether

The influential wave theory of light proposed by Augustin Jean Fresnel (1788–1827) made a major contribution to the abandonment of imponderable fluid theories. The origins of Fresnel's work lay in his opposition to the scheme of imponderable fluids and to the Laplacian corpuscular theory of light and the caloric theory of heat. At the time of his first interest in optics in 1814, Fresnel wrote that he suspected that light and heat were connected with the vibrations of a fluid. His commitment to the concept of light as a form of motion of a medium was basic to his optical theory. By 1821 he had reformulated the science of optics in terms of the dynamics of a wave-propagating medium, the luminiferous ether. Fresnel rejected the corpuscular theory of light as implying that light corpuscles were an anomalous form of matter, an imponderable fluid; in his theory the vibrations of the ether explained the phenomena of optics. Repudiating a disjunction between ordinary matter and imponderable substances, Fresnel envisaged a unified physics based on the mechanical properties of the ether, conceived as a form of ordinary matter. He abandoned the corpuscular 'emission' theory of light in favour of the theory of the propagation of light as undulations in the luminiferous ether.

Fresnel became familiar with the complexity of contemporary corpuscular optics only as a result of meeting Arago in 1815; still unfamiliar with Young's work, he proposed an explanation of optical diffraction by means of the principle of interference. He supported his interpretation of diffraction bands as the constructive and destructive interference of waves by an experimental investigation, finding a close correspondence between the observed and predicted positions of the diffraction bands. His paper on optical diffraction, submitted for the Paris Academy prize competition in 1819, provided a mathematical theory of the interference of light

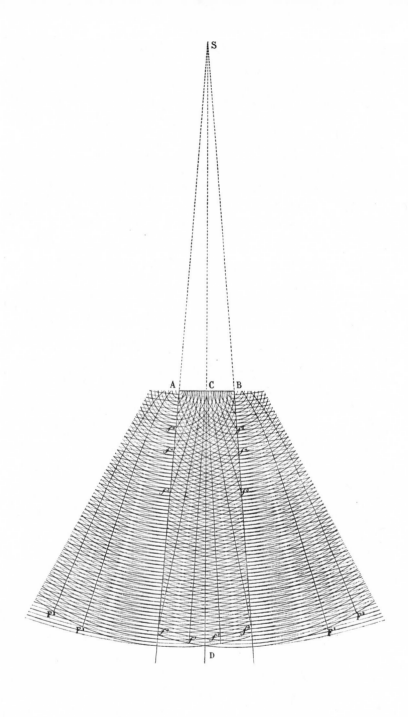

waves. Although Fresnel won the prize – his theory received unexpected experimental confirmation during the judging – his paper did not lead to a dramatic conversion of the Laplacians to the undulatory theory of light.

The phenomenon of the polarisation of reflected light discovered by Malus in 1808 implied that light possessed asymmetric properties, and this presented a major problem to Fresnel's theory. While Malus and Biot explained the polarisation of light in terms of the asymmetric motions of light particles, Fresnel's concept of waves propagated in a fluid ether could not have the asymmetric properties implied by the polarisation effect. Fresnel realised that

---

Fig. 2.2. Diffraction bands discussed by Fresnel in his first paper on the diffraction of light (1816). Fresnel studied the diffraction of light when a hair or other thin object was illuminated by a narrow beam of light, observing alternating bands of light and dark. He interpreted this effect by appeal to the supposition that light was undulatory and to the principle of the interference of light waves. Bright bands occurred where the light waves reflected by the diffracter (AB) and those proceeding directly from the light source (S) were in phase and would reinforce one another, whereas the intervening dark bands occurred when the vibrations constituting light from the light source and those reflected by the diffracter were out of phase and would therefore cancel one another.

Fresnel was able to demonstrate a close agreement between the positions of the diffraction bands predicted by his theoretical treatment of the interference of light waves and the values obtained by an experimental test. His explanation of diffraction bands by the constructive and destructive interference of light waves provided an important argument in favour of the wave theory of light, an interpretation that received strong support when he confirmed his theoretical predictions experimentally. Arago, who had already signalled his support for the wave theory of light, reported favourably on Fresnel's work and fostered Fresnel's further research. Biot responded by giving a corpuscular interpretation of diffraction, and the supporters of Laplacian corpuscular optics proposed a prize contest on the subject, seeking a corpuscular explanation of diffraction and the refutation of the wave theory of light. Fresnel's successful paper, submitted to the Paris Academy in 1819, won an important victory for the wave theory of light. During the judging, Poisson argued that Fresnel's theory implied the unexpected consequence that the centre of the shadow of a disk used as a diffracter would be illuminated, a result that was confirmed by experiment. Source: Oeuvres complètes d'Augustin Fresnel, ed. H. de Senarmot, E. Verdet, and L. Fresnel, 3 vols. (Paris, 1866–70), 1:23.

the analogy between light and sound waves, which had fruitfully suggested the concept of interference, was false. If the undulatory theory of light was correct, a more searching investigation of the mechanical properties of the ether was required, for the asymmetric property of polarised light was incompatible with Fresnel's assumption that light waves were longitudinal vibrations in a medium (like sound waves), vibrating in the direction of the propagated wave. To explain polarisation, Young proposed what he termed an 'imperfect explanation': That light consisted of a 'transverse vibration' like the undulations of a plucked cord, vibrations at right angles to the direction of the propagated wave.

To explain the polarisation of light, Fresnel considered the hypothesis that the vibrations constituting light consisted of two components at right angles to one another, a combination of longitudinal and transverse vibrations, supposing that polarisation destroyed the longitudinal component. By 1821, as a result of careful work on the nature of optical polarisation, he realised that the vibrations constituting light were purely transverse. He proposed a model of an ether consisting of molecules bound by forces acting at a distance as the medium in which the transverse vibrations were propagated. In reply to criticisms by Poisson that transverse waves would not be satisfactorily transmitted in a fluid medium, Fresnel argued that the problem lay in the reconstruction of the theory of the ether rather than in the futile attempt to deny the wave theory of light. Fresnel realised that a fluid ether could not produce transverse vibrations; to give rise to transverse vibrations the ether would have to possess the property of rigidity. The elaboration of a mechanical model of the ether had not been his primary intention, and was undertaken only in support of his undulatory theory of light. Fresnel's conclusion that light waves were transverse vibrations posed the problem of constructing a mechanical model of the ether capable of transmitting transverse rather than longitudinal waves, a problem that became a major feature of nineteenth-century optics.

Fresnel's work was of major importance for the development of physics in the nineteenth century, providing as it did the paradigm for the unification of the phenomena of physics in the construction of a mechanical model. The construction of such models came to be a central feature of physical theorising. The laws of optics and, by implication, those of other physical phenomena could be brought within the framework of mechanical explanation and rendered intelligible by reference to the mechanics of solid and fluid media.

The Luminiferous Ether and Mechanical Explanation

Fresnel's work inaugurated a series of studies of the physical properties of the ether intended to establish the mathematical basis of the principles of optics and the precise mechanical structure of the ether. In a major paper on the theory of light published in 1830, Augustin Louis Cauchy (1789–1857) formulated a mathematical treatment of the principles of optics that took as a premise an ether with the mechanical properties of an elastic solid medium. Cauchy's major innovation was to demonstrate that the propagation of transverse vibrations of light could be obtained from the differential equations of motion of an elastic solid. The difficulty with his formulation arose from his failure to justify his assumptions about the molecular structure of the elastic solid ether; nor could he justify the boundary conditions he imposed so as to derive the laws of optics. Although Cauchy's work was mathematically more sophisticated than Fresnel's, it raised similar difficulties: The mechanical structure of Cauchy's elastic solid ether was open to question and its relation to the principles of optics obscure.

Cauchy's work was especially influential in Britain, where it stimulated investigations into the mechanical foundations of optics and the structure of the ether. A new generation of mathematicians and physicists, including John Herschel (1792–1871), William Whewell (1794–1866), and George Biddell Airy (1801–92), all of whom supported the wave theory of light in the 1830s, had introduced Continental analytical mathematics into Cambridge mathematics teaching. The wave theory of light, especially in the form elaborated by Cauchy, in which optical laws were deduced from the differential equations of motion of an elastic solid, exemplified their belief in the application of mathematics to physical problems. Despite objections from supporters of the Newtonian 'emission' theory of light that some of the predictions of the wave theory had not been confirmed experimentally, the supporters of the wave theory stressed its mathematical sophistication and defended the hypothesis of the luminiferous ether as a valid basis for optical theory.

In an important paper published in 1838, George Green (1793–1841) attempted to provide a secure mechanical foundation for the theory of the elastic solid ether, deriving the laws of optics from a mathematical function that expressed the mechanical properties of the ether in an analytical form. Criticising Cauchy's assumptions about the molecular structure of the ether, Green emphasised the gap between physical reality and any model employed to represent reality, and the difficulty of specifying the structure of the luminiferous ether. He argued that it was preferable to assume a general 'physical principle as the basis of our reasoning, rather than assume certain modes of action, which, after all, may be widely different from the mechanism employed by nature'. Green's paper was of great importance in offering as an alternative to the construction of specific mechanical models an explanation using analytical dynamics. This mode of explanation, termed a 'dynamical' theory by British physicists, was based on the formalism of analytical dynamics employed by Joseph Louis Lagrange (1736–1813) in his *Mécanique analytique* [*Analytical mechanics*] (1788), and was not linked to any specific physical model; it provided a justification of the intelligibility of mechanical explanations of phenomena.

A similar approach to the problem of justifying the mechanical theory of the elastic solid ether was adopted by James MacCullagh (1809–47). In his 'Essay towards a dynamical theory of crystalline reflexion and refraction', written in 1839, MacCullagh sought to establish links between the principles of optics and mechanics, by using the methods of Lagrangian analytical dynamics to derive the laws of optics. MacCullagh argued that the elasticity of the ether did not depend on the distortion or compression of the ether, but only on the rotational displacement of its elements, and he derived the laws of the propagation of light from an analytical function that was dependent on the rotational elasticity of the ether. Emphasising that 'the constitution of the luminiferous medium is entirely unknown', MacCullagh urged his 'dynamical theory' of light as the only appropriate mode of theory construction; yet he admitted that his justification of rotational elasticity (without reference to a mechanical model) 'cannot be regarded as sufficient, in a mechanical point of view'. MacCullagh's work remained controversial, and his theory of the rotationally elastic ether came to have an important influence only when, as a result of the dissemination of Maxwell's electromagnetic theory of light in the 1870s, it was argued that rotation was a fundamental mechanical property of the ether.

An alternative approach to the elaboration of a mechanical theory of the elastic solid ether was followed by George Gabriel Stokes (1819–1903) in a series of papers published in the 1840s. While avoiding speculation about the molecular structure of the ether, and therefore seeking to avoid the difficulties apparent in Cauchy's work, Stokes stressed the physical structure of his ether model. He formulated a mathematical theory of an ideal continuous solid analogous to Euler's theory of the mechanics of fluid media, in which the equations of motion of a continuous fluid were derived without reference to its molecular structure. Though Stokes supposed that the ether was particulate in structure, he avoided using a molecular hypothesis to explain the propagation of light through the ether. He suggested that the ether might act like a fluid with respect to the motion of the earth and planets through it, but like an elastic solid with respect to the vibrations constituting light. He represented the ether as analogous to a glue–water jelly; a mixture containing little glue would behave like a fluid for the translational motion of large bodies through it, but would still possess elasticity and hence produce small transverse vibrations corresponding to the transmission of light. Stokes thus avoided a theory of the molecular structure of the ether, but employed his mechanical glue–water model to represent its elasticity.

The work on ether dynamics initiated by Fresnel thus had an important influence in fostering the programme of the mechanical theory of nature; and the theories of MacCullagh, Green, and Stokes offered contrasting approaches to the problem of the mechanical explanation of physical phenomena. Stokes's theory of the ether involved the postulation of a mechanical representation, an appeal to a mechanical analogue drawn from ordinary experience, whereas MacCullagh and Green appealed to mechanical explanation by demonstrating the application of analytical, generalised equations of motion. Physicists did not always view these two approaches to theory construction as contradictory alternatives, but sometimes saw them as compatible and even complementary procedures. Whereas the limitations of pictorial models were appreciated, the abstract character of the Lagrangian dynamical formalism was frequently seen as insufficiently realistic. The construction of mechanical explanations, based on physical models or Lagrangian dynamical theory, became a distinctive feature of British physics, subsequently applied to the representation of electromagnetism by William Thomson and by Maxwell.

Fourier and the Mathematical Methods of Physics

In his *Théorie analytique de la chaleur* [*Analytical theory of heat*] (1822), Jean Baptiste Joseph Fourier (1768–1830) made a contribution of fundamental importance to the creation of a unified physics grounded on mathematical principles. Concerned with the mathematical theory of heat, Fourier did not so much reject the theory of heat as an imponderable fluid (caloric), responsible for the repulsive force that separated the particles of ordinary matter, as avoid all questions about the physical nature of heat. Fourier set the study of heat in the tradition of rational mechanics, basing it on differential equations that characterised the transmission of heat, equations that were independent of all physical hypotheses.

Fourier presented his theory of heat as methodologically analogous to Newton's mathematical theory of gravity: The cause of gravity was unknown, but its effects could be discovered by observation and subjected to mathematical analysis. Fourier envisaged the analytical theory of heat as an extension of the conceptual range of rational mechanics. He sketched the history of mathematical mechanics from Archimedes, Galileo, and Newton to the achievements of the geometers of the eighteenth century, noting that this tradition had confined its work to the study of mechanics, to the problems of dynamics and statics and the study of the principles of motion and equilibrium of solid and fluid bodies. The range of these mechanical theories did not apply to the effects of heat, which comprised a special set of phenomena. Nevertheless, the analytical equations employed in the study of the mechanical properties of bodies could be applied to a wider range of phenomena. Fourier declared that 'mathematical analysis is as extensive as nature itself'; the clarity and universality of mathematical language enabled all phenomena to be subsumed under mathematical laws, revealing the unity and harmony of natural laws.

In considering the mathematical theory of heat within the framework of rational mechanics, Fourier avoided hypotheses about the nature of heat. Just as rational mechanics was based on observable quantities, Fourier's mathematics was based on the effects of heat, not its hypothetical cause; on the temperature distribution in bodies, not on the way in which the repulsive power of heat determined the physical states of material substances. Fourier declared that the basic principles of the theory were derived from 'a very small number of primary facts' which were then subjected to a mathematical analysis – a method analogous to the procedures of rational mechanics.

Fourier's work on heat originated in a study of the transmission of heat between discrete particles; probably stimulated by a paper by Biot on the flow of heat in a metal bar, he turned to an investigation of the flow of heat in solid bodies, presenting his work to the French Institute in 1807. Expanded in 1811 to include the treatment of the diffusion of heat in infinite bodies, Fourier's revision of his paper won the prize competition on heat diffusion set by the institute, despite objections raised by Lagrange to some of Fourier's mathematical arguments. Fourier's work also contra-

equation could therefore be applied to the theory of electricity, and the continuity or conservation of the flux of heat implied the conservation of the flow of electric force. The unification of the phenomena implied by the continuity equation was mathematical rather than physical, a unity of geometrical form rather than a physical analogy between heat and electricity.

Fourier's work highlighted the importance of a mathematical formalism that was independent of a theory about the constitution of matter, pointing up a distinction between mathematical and physical representation. Though nineteenth-century physicists did not eschew physical models, the construction of models was distinguished from the representation of physical reality. As Green observed in presenting his mechanical theory of the ether, there was a distinction between the supposition of a model and the 'mechanism employed by nature'.

The Electrical Ether

The theory of the luminiferous ether provided a physical model for the propagation of light through space in an ambient medium. The discovery of electromagnetism in 1820 by the Danish physicist Hans Christian Oersted (1777–1851) led to the elaboration of a theory of the propagation of electric forces in an electrical ether. The concept of electrical 'atmospheres' surrounding particles of ordinary matter was common in the writings of eighteenth-century physical theorists, but by the late eighteenth century it had been transformed into the concept of the 'sphere of activity', the space in which electrical forces were manifested. Many electrical theorists ascribed electrical action to stresses in an ethereal medium, discussing the physical basis of the spatial distribution of electricity. The relations between electricity and magnetism and the possible connections between the electric and magnetic fluids were frequently discussed, but although the analogy between electricity and magnetism was recognised, they were conceived as independent phenomena.

Oersted's discovery of the action of an electric current on a magnetic needle transformed electrical science, emphasising the unity of electrical and magnetic phenomena. The discovery of 'electromagnetism' was the culmination of Oersted's research on electricity, in which he was guided by his conviction of the underlying unity of the powers of nature. He was committed to a view of nature that envisaged electricity, magnetism, heat, and light as manifestations of a single power, and he sought to explain

dicted the prevailing Laplacian style of theory construction, and Poisson continued to oppose Fourier's analytical mechanics and to urge the physical mechanics of Laplacian physics. In his own theory of heat Poisson employed the physical model of caloric fluid radiated directly to a distance. Fourier's mathematical methods were quite different, focusing on the expression of the differential equation of the diffusion of heat, rather than postulating a physical model to represent heat flow. He based his treatment of the transmission of heat on the distribution of temperature in solid and liquid bodies, arguing that the transmission of heat could be expressed as the communication of heat between molecules. The heat transmitted between two molecules of a substance was proportional to the difference in their temperature and a function of their distance apart, this function varying with the nature of the substance. The flux of heat was related to the temperature difference; assuming that heat was conserved during its flow, Fourier obtained the 'continuity equation' (derived by Euler for the flow of a fluid), a differential equation relating the flow of heat and the temperature gradient. Formally identical to Poisson's results, this equation was not in Fourier's analysis based on a physical model; the nature of heat was not considered. Fourier aimed to reduce physical problems to questions of mathematical analysis. The theory was based on empirical laws of the distribution of temperature that could be tested by experiment. The equations of the propagation of heat were derived from the basic facts of temperature distribution; whatever the nature of heat, its propagation could be represented by these differential equations.

Fourier's analytical theory of heat had important implications for subsequent mathematical physics. He had broadened the framework of rational mechanics to include the problem of the propagation of heat, and this provided a paradigm for a mathematical physics not confined to narrowly mechanical problems. Fourier's method also stressed the primacy of mathematical laws and the distinction between mathematical theory and its physical interpretation. This distinction was important in William Thomson's 1842 elaboration of a theory of electrostatics, which employed a mathematical formalism analogous to the mathematical theory of heat distribution of Fourier's analytical theory of heat, expressing an analogy between thermal conduction and electrostatic attraction. Thomson proposed a geometrical model, common to both electrostatics and thermal conduction, in which the distribution of electricity was represented by a flux of electrical force and the distribution of heat was represented by a flux of heat. The continuity

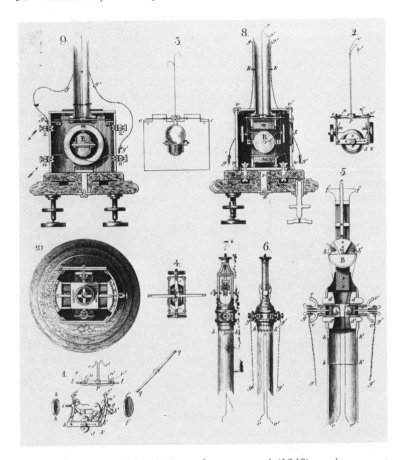

Fig. 2.3. Wilhelm Weber's 'electrodynamometer' (1848), an instrument developed for the precise measurement of electrical forces. In the electrodynamometer a coil of copper wire was suspended within a larger fixed coil. When a current passed through the wires of both coils the suspended coil rotated, and the movement was counteracted by the suspension of the coil; the rotation of the inner coil provided a measure of the electric current. The figure displays the intricate construction of the instrument so as to allow the free rotation of the coil and hence permit the precise measurement of small electric currents; it illustrates a close collaboration between physicist and instrument maker. The concern with quantitative measurement was fundamental to Weber's approach to physics. He emphasised that the most rigorous standard of experimental precision was required, and he regarded the traditional qualitative approach to experimentation as obsolete.

natural phenomena in terms of a conflict between the forces of attraction and repulsion. These ideas were commonplace in late eighteenth-century physics; but Oersted was influenced by their expression in the German *Naturphilosophie* (nature philosophy) tradition, which stressed the unity of nature and the polarity of forces, and by Kant's theory that fundamental forces of attraction and repulsion were the defining properties of matter. Oersted considered an electric current to be a dynamic oscillation, an undulation of forces produced by a conflict between opposing attractive and repulsive forces. He observed that the magnetic needle was deflected in opposite directions when placed above and below the wire carrying the electric current, and he inferred that the undulating electric powers in the wire produced a circular motion in the space surrounding the wire. Oersted's experiment implied that electric and magnetic forces acted in circles and were spatially distributed.

Oersted's discovery prompted a variety of hypotheses to explain the spatial disposition of electromagnetic interactions. Physicists depicted the forces geometrically by magnetic curves filling space, or represented the circulating magnetic forces by magnetic fluid vortices in an ambient ether. An alternative hypothesis, proposed by André Marie Ampère (1775–1836), employed an analogy with the propagation of light waves to explain the propagation of electromagnetic action. Ampère extended Oersted's results by demonstrating the interactions between current-carrying wires, arguing that magnetism could be explained by electric currents, and he elaborated a mathematical theory of electromagnetism based on attractive and repulsive forces between infinitesimal current elements. Ampère explained the propagation of electromagnetic action by drawing on Fresnel's concept of the luminiferous ether. He argued that the ether was constituted of positive and negative electric fluids; electromagnetic phenomena arose from the disturbance of the electric fluids, whereas light was produced as a result of the vibrations of the fluid. The luminiferous and electromagnetic space-pervading ether provided a physical model for the propagation of electromagnetic action, and by the 1820s the concept of an ether propagating light and mediating electromagnetic interactions was established.

Oersted's discovery also prompted the search for other electromagnetic effects, a search that led to the discovery of electromagnetic induction by Michael Faraday (1791–1867) in 1831. Faraday found that the passage of an electric current through a wire (the primary circuit) wrapped round one side of an iron ring induced a transient current in a wire (the secondary circuit) wound round the other side of the ring. To explain the transient induction of electricity in the secondary circuit when the primary circuit was closed, Faraday suggested that the effect was due to the propagation of a 'wave of electricity'. He argued that electric and magnetic actions were propagated in time, and that they were to be considered as resulting from a progressive motion; he compared the diffusion of magnetic forces to the transmission of light and sound waves in an ambient medium. Faraday was familiar with the speculations of Oersted and Ampère on the propagation of electromagnetic forces and probably also with the earlier theories of the agency of an ethereal medium as the vehicle for electrical action. These theories of the spatial disposition of electrical forces provided the framework of ideas in which he interpreted his discovery of electromagnetic induction, and the context in which he was subsequently to emphasise the spatial distribution of electrical and magnetic forces in elaborating his theory of the electromagnetic 'field'.

Conversion Processes and the Unity of Nature

The belief in the interconversion of natural powers and the unity of nature, grounded on an awareness of the relationships and connections among heat, light, electricity, and chemistry, was common in late eighteenth-century physics, and was frequently based on the theory that the various imponderable fluids were modifications of a repellent material substance distinct from ordinary matter. The concept of the unity of nature attained especial significance in the 1830s. The relationships between chemical and electrical forces explored by Davy, the discovery of electromagnetism by Oersted

---

Fig. 2.3 (*cont.*)

This approach reflects the professionalisation of physics in early nineteenth-century Germany, which involved a stress on the value of research, an emphasis on the search for mathematical laws and precise experimentation, and the creation of research and teaching laboratories. These ideals shaped the development of the discipline of physics in the nineteenth century and were especially important in Germany, where research and specialisation became the norms of physics, and the professionalisation of physics was fostered by the state funding of university research and state direction of university appointments. *Source*: Wilhelm Weber, 'Elektrodynamische Maassbestimmungen', *Annalen der Physik* 73 (1848), 336.

and of electromagnetic induction by Faraday, and the establishment of the analogy between light and radiant heat by Macedonio Melloni (1798–1854) were important in establishing the implication of conversion processes, the equivalence of different natural phenomena. The concept of the unity and conversion of natural powers formed one of the strands that, transformed, became articulated by around 1850 as the principle of the conservation of energy. The experimental discoveries of Oersted, Faraday, and Melloni in the 1820s and 1830s did not in themselves lead to an awareness among physicists of the unity of nature and the conversion of natural powers; this doctrine had long been familiar. In abandoning the imponderable fluid theories, physicists retained the concept of the unity and conversion of natural phenomena. Divorced from the imponderable fluid theories and reinforced by the new experimental discoveries of conversion phenomena, the doctrine of the unity and equivalence of natural powers led in the 1830s to a heightened awareness of the relationships among natural phenomena.

The development of the concept of the luminiferous ether provided a new context for discussion of the unity of nature. Fresnel had conceived the possibility of explaining light, heat, and electricity as the modifications of a universal ether, and Ampère had formulated the concept of an electrical ether as the vehicle for electromagnetic and optical actions. In the 1830s Ampère developed a comprehensive theory of the analogy between light and heat. He argued that heat was not the result of the motion of an imponderable fluid but was caused by a vibratory motion of the molecules of bodies; the vibrating molecules did not transmit their motions directly, but acted through the mediation of the ether. Heat was therefore propagated by the undulations of the ether. Ampère's 'wave theory' of heat received considerable support; the conventional assertion of the fundamental identity of heat, light, and electricity was seen in a new context, that of the dynamics of the ether, whose motions were responsible for the transmission of these phenomena. The dynamics of the ether rather than the behaviour of imponderable fluids became the basis for a unified physics.

The emphasis on the unity and conversion of natural powers or 'forces' was fundamental to the scope and direction of Faraday's experimental researches. His electrochemical experiments in the 1830s were designed to unravel the relationships between chemistry and electricity, and he affirmed the identity of the forces of chemical affinity and electricity. He drew attention to the interconversion of the powers of heat, electricity, and chemical affinity as a result of their essential unity. In his view the conversion of natural powers was a consequence of their indestructibility: Natural powers could not be created from nothing, and the generation of one power was consequent on the exhaustion of another. Much of Faraday's work was concerned with conversion phenomena: electromagnetic induction, electrochemistry, and the action of magnetism of light. He also speculated on the possible conversion of electricity and gravity, striving to demonstrate their connection by experiment. In Faraday's physics the idea of the conversion of natural powers was detached from the theory of imponderable fluids. He stated his fundamental belief in the unity and equivalence of 'forces' in his paper on the action of magnetism on light in 1845: 'The various forms under which the forces of matter are made manifest have one common origin; or, in other words, are so directly related and mutually dependent, that they are convertible, as it were, one into another, and possess equivalents of power in their action'.

The concept of the conversion of forces was central to Faraday's physical world view, providing the conceptual framework that made his discoveries intelligible and defining the nature of his experimental investigations. This doctrine was given a popular exposition by William Robert Grove (1811–96) in his *Correlation of physical forces* (1846), often interpreted subsequently, and retrospectively, as a statement of the principle of the conservation of energy. Asserting the indestructibility and convertibility of natural powers or forces, Grove sought to demonstrate the implications of conversion phenomena. By the 1840s these ideas were part of the fabric of physical theory.

The Unity of Nature: Heat and Mechanical Work

The statement of the interconversion and equivalence of mechanical 'work' and heat was a key feature in the establishment of the principle of the conservation of energy. In arguing that apparent losses of mechanical energy in mechanical processes were to be equated with the generation of heat, the principle of energy conservation asserted the unity of natural phenomena. Eighteenth-century physicists had considered energy losses within mechanical systems to be isolated from nonmechanical processes, and therefore they did not enunciate a theory of the equivalence of heat and mechanical energy. The concept of the conservation of mechanical energy was, however, to be found in eighteenth-century treatises on mechanics; it was familiar, in the form first enunciated by Leibniz, as the principle of the conservation of *vis viva* ('living force'). Leibniz had argued that living force, measured by the product of the mass and the square of the velocity, was conserved in mechanical processes. Hence the activity of the universe was conserved; nature was not like a clock that required rewinding.

The use of the concept of the conservation of living force by eighteenth-century physicists did not entail a commitment to Leibniz's theory of nature. The principle of the conservation of living forces came to be widely used, especially in discussing the collision of perfectly elastic bodies. Johann Bernoulli provided the most systematic and penetrating expositions of the living force concept in papers published in the 1720s and 1730s. He and Leibniz had recognised that there would be apparent losses of living forces in the collision of inelastic bodies, and Bernoulli argued that inelastic bodies were analogous to springs that were impeded from expanding after being compressed. Thus living force would be consumed in the compression of bodies, but was not destroyed in their deformation. Although Daniel Bernoulli later discussed the operation of a heat engine, which used the living force stored in coal by the generation of gases from coal, he did not suggest the equivalence of heat and work or attribute the mechanical losses of living force in inelastic collisions to heat; he conceived such losses in strictly mechanical terms. Writing around 1760, Cavendish applied the principle of the conservation of living force to the communication of heat: Supposing that heat was produced by the vibration of the particles of bodies, the transmission of heat was determined by the conservation of the living forces of the particles of bodies. Although he followed Johann Bernoulli in arguing that living force was not destroyed in inelastic collisions, Cavendish did not equate the losses of living force with the generation of heat. He considered mechanical and thermal systems to be analogous but separate.

By the early nineteenth century the measure of living force by mechanical 'work', the product of force times distance (also referred to as 'mechanical power' or 'mechanical effect'), had been introduced into writings on power technology. The operation of heat engines that generated work by burning coal suggested the conversion of work and heat; and Peter Ewart (1767–1842) implied the quantitative relation between the heat generated by the burning of coal and the 'mechanical power' or 'force' it could be employed to produce. By the 1820s theoretical treatises on mechanics stressed the concept of 'work', defined it as measured by the integral of force with respect to distance, and clarified its mathematical relationship to the concept of living force, showing that it provided a measure of mechanical 'energy'. The concept of work as a measure of mechanical energy established a quantitative basis for conversion processes.

By the 1840s the quantitative equivalence of mechanical work and heat had been made explicit by several physicists and engineers, who calculated the conversion coefficient of heat and work, the 'mechanical equivalent' or 'mechanical value' of heat. The clarification of the concept of mechanical work and the demonstration of the quantitative equivalence of heat and work were important factors in the enunciation, experimental demonstration, and mathematical conceptualisation of the principle of the conservation of energy. The assertion of the equivalence of work and heat brought mechanical processes explicitly into the network of conversion processes discussed by physicists. The law of the conservation of energy provided a conceptualisation of this framework of explanation: a universal principle of the interconversion of natural powers, together with a quantitative measure of conserved physical quantities. The establishment of the law of energy conservation was also associated with the enunciation of the mechanical theory of heat, the theory that heat was the result of the vibration of the particles of bodies, which provided an explanatory basis for the interconversion and equivalence of mechanical work and heat. A mechanistic ontology of matter in motion provided an explanatory foundation for the conversion of natural powers and the conservation of energy.

The work of James Prescott Joule (1818–89) in the 1840s was of fundamental importance in tracing out the network of conversion processes, and in providing experimental confirmation of the quantitative equivalence of heat and mechanical work. Joule's statement of the mechanical theory of heat provided a connecting link between the principle of the conservation of energy and the programme of mechanical explanation. In the early 1840s Joule's research had focused on the improvement of electrical engines, and also on electrochemistry – an interest shaped by Faraday's emphasis on the electrical basis of chemical reactions. Davy and Faraday had sought to formulate an electrical theory of chemical affinities, and Joule attempted to elaborate this electrical theory, seeking to unify electrical, chemical, and thermal phenomena by demonstrating their interconversion and quantitative equivalence. To measure the quantitative relation between mechanical work and heat, he constructed an electrical engine in which mechanical work generated

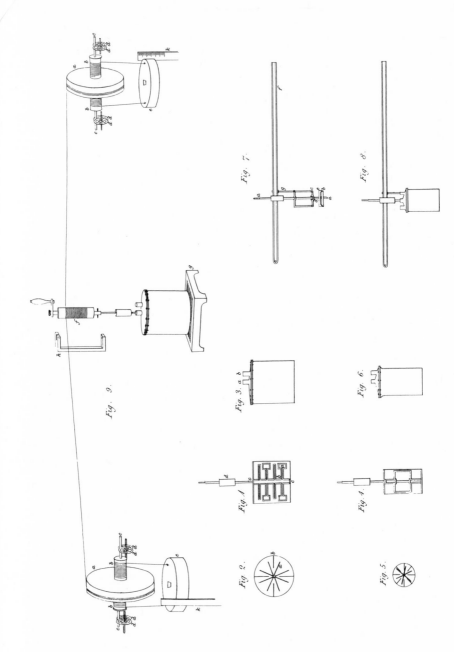

Fig. 9.

Fig. 7.

Fig. 8.

Fig. 3. a b

Fig. 6.

Fig. 1.

Fig. 4.

Fig. 2.

Fig. 5.

Fig. 2.4. J. P. Joule's paddle-wheel apparatus, which he employed to establish the mechanical equivalent of heat (1850). Joule's Royal Society paper was the culmination of his 1840s research on the relation between heat and work; he used a method (the generation of heat by fluid friction with a paddle-wheel apparatus) that he had already employed, but obtained a more accurate measurement of the mechanical equivalent of heat. Figures 1 and 2 represent the horizontal and vertical plan of the brass paddle wheel, showing the arrangement of rotary arms and stationary vanes (to keep the liquid stable). Figure 3 represents the copper vessel containing water into which the paddle wheel was fitted; it had a second neck for the insertion of the thermometer to measure the increase in water temperature. Figure 9 shows the whole apparatus, including the weight and pulley machinery used to set the paddle wheel in motion. Figures 4, 5, and 6 show the modified apparatus (consisting of a wrought-iron paddle wheel and cast-iron vessel) used for experiments on the friction of mercury. Figures 7 and 8 illustrate the apparatus employed for the friction of solids. By measuring the distance the weights descended, Joule was able to establish the relationship between the work performed and the heat generated. He was careful to allow for the effects of the conduction and radiation of heat, and he claimed to read his thermometers to an accuracy of one two-hundredth of a degree Fahrenheit.

Joule stated two conclusions: that the quantity of heat generated by friction was proportional to the quantity of work expended, and that the quantity of heat to raise one pound of water one degree Fahrenheit required the expenditure of mechanical work equivalent to the fall of 772 pounds through the space of one foot. A third conclusion, that friction consisted in the conversion of work into heat, was omitted, suppressed at the suggestion of the Royal Society referee of the paper (probably Faraday). The referee argued that the conclusion demonstrated by Joule, that the generation of a given amount of heat always required the expenditure of the same quantity of work, did not entail the consequence that heat was *convertible* into work and work into heat. *Source*: James Prescott Joule, 'On the mechanical equivalent of heat', *Philosophical Transactions of the Royal Society* 140 (1850), 64.

an electric current, which in turn generated heat; the mechanism enabled the numerical relation between heat and mechanical work to be evaluated. Joule soon concluded that mechanical work could be directly transformed into heat by friction; he maintained that 'wherever mechanical force [work] is expended, an exact equivalent of heat is *always* obtained', supporting his statement of the indestructibility and self-sufficiency of natural powers by an appeal to the argument that only God could destroy the agents of nature.

The determination of the 'mechanical value' of heat provided a quantitative measure of conversion processes, and in Joule's view the connection between heat and 'mechanical power' implied a theory of heat as a mode of motion. Joule's electrical theory of nature helped shape his speculations on the nature of heat. He suggested that the velocity of rotating electrical atmospheres surrounding atoms would determine temperature, and elaborated his electrical theory of chemical affinity to explain the nature of the motion that constituted heat. However, the focus of Joule's theory shifted from the emphasis on electricity as the mediating agent for the conversion of heat into mechanical work. His demonstration that mechanical work was directly transformed into heat by friction and his measurement of the mechanical value of heat in an experiment to test the heat produced by friction led to a statement of the interconversion, indestructibility, and quantitative equivalence of all natural powers. The transformations of mechanical power, rather than the mediating role of electricity in the interconversion of heat and work, became the primary concept in his theory of nature. By 1847 he illustrated the direct relation between mechanical work and heat by sketching cords linking a weight to rotating atoms, so that the weight could be raised or lowered, corresponding to a decrease or increase in temperature. Joule did not publish his sketch of the motion constituting heat, merely stating that heat was measured by living force and hence that the particles of heated bodies were in motion.

Joule formulated a general physical theory of the convertibility of heat and work that brought the concept of the mechanical equivalent of heat within the framework of the doctrine of the unity and conversion of natural powers. He argued that it was by means of the conversion of 'forces' that order was maintained in the universe, and that the indestructibility of forces demonstrated the self-sufficiency of nature. Once God had established the framework of natural powers, these forces would remain constant in their total effect. Joule had not begun his research with theoretical presuppositions relating to the connections among different natural powers, and he employed the idea of the conversion and indestructibility of natural powers as a framework for the theoretical interpretation of his experimental discoveries. Although this theory did not directly regulate his research, it provided an intellectual context for his experimental demonstrations of conversion phenomena. Joule maintained that he had established the mutual convertibility of heat and work; he did not claim the formulation of a general principle of the conservation of 'energy'. As he told William Thomson in 1848, he had sought to provide a 'proof of the convertibility of heat into [mechanical] power'. Joule thus affirmed the 'possibility of *converting heat into mechanical effect* [work]', a doctrine which he believed was fundamental to the theory of the steam engine. The physical intelligibility of the quantitative equivalence and mutual convertibility of heat and work was supported by the assertion of the mechanical theory of heat: that heat consisted of the motion of the particles of matter.

Helmholtz and the Conservation of Energy

In writing his seminal memoir *Ueber die Erhaltung der Kraft* [*On the conservation of force*] (1847), Hermann von Helmholtz (1821–94) provided a mathematical formulation of the principle of conservation of energy. Helmholtz's concept of the 'conservation of force' retained a suggestive ambiguity, denoting the indestructibility and transformability of natural powers or forces as well as the conservation of energy, subsuming the conversion of natural powers under the framework of the conservation of energy, and in this way giving a mathematically precise designation of the quantities that were conserved.

The origins of Helmholtz's energy theory lie in his interest in physiology and his specific concern with the problem of animal heat. A leading figure in the Berlin school of physiologists, who sought to base physiology on physical principles, Helmholtz attempted to demonstrate that the body heat and muscular action produced by animals could be derived from the oxidation of foodstuffs. Helmholtz was profoundly influenced by the work of Justus von Liebig (1803–73), who had attempted to derive physiological phenomena from physical and chemical laws. Despite the ambiguity of the experimental evidence, Helmholtz concluded in favour of Liebig's theory that respiration was the only source of animal heat, noting that Liebig's argument was dependent on the validity of the principle of the 'constancy of force', the principle of the indestructibility and transformability of forces. Helmholtz set himself the task of justifying the principle that natural powers or forces could be transformed into one another but could not be annihilated.

Though he attacked the view that organisms possessed forces entirely peculiar to living things and unlike the forces operative in the physical realm, he correctly interpreted Liebig's approach to physiology as one that held organic life to be the result of forces that were modifications of those operative in inorganic nature. Helmholtz did not reject Liebig's use of the concept of vital forces. The crucial problem for Helmholtz was the explanation of the forces that regulated the physiology of organisms by the laws determining inorganic forces, and he maintained that this explanation required the operation of all forces to be subject to the law of the constancy of force. Because Liebig's vital forces did have the same character as inorganic forces in that the vital force could not arise from nothing and would generate an equivalent amount of another force, Helmholtz maintained that Liebig's theory depended on the assumption of the constancy of force. Helmholtz did, however, have one objection to a vital force principle: A vital force could be regarded as self-perpetuating and hence could not be subject to the principle of the constancy of force.

The denial of perpetual motion was fundamental to Helmholtz's commitment to the principle of the constancy of force. His purpose in writing the paper was to demonstrate the principle of *Erhaltung der Kraft* ('conservation of force') by a mathematical investigation of the physical quantities that were conserved. In this investigation the concept of conservation or constancy of force acquired a sense additional to that in which it denoted the indestructibility and transformability of natural powers: that of the conservation of energy. Helmholtz's usage thus embraced the idea of the conversion of forces (as employed by Faraday and Joule) and the mathematically precise law of the conservation of energy, a law that, as Helmholtz observed after the energy concept had become current in the 1850s, gave a more exact description of the conserved quantities. Helmholtz sought to demonstrate that the conservation of energy could be expressed in an explanatory framework that posited an ontology of matter and of the forces associated with material bodies – a mechanistic ontology of matter in motion. Helmholtz thus used the term 'force' in yet a third sense in his paper: to denote Newtonian central forces of attraction and repulsion; he maintained that 'the problem of physical science is to reduce natural phenomena to unalterable forces of attraction and repulsion, whose intensity depends on the distance'. His formulation of the principle of the conservation of energy was therefore linked to his assumption of the universal validity of the mechanical theory of nature, which he declared to be 'the condition of the complete comprehensibility of nature'.

In support of his contentions that the intelligibility of nature required that nature be considered as regulated by causal laws, and that these laws were Newtonian central force laws, Helmholtz appealed to Kant's metaphysics of nature. Whereas Kant's discussion of matter in terms of forces of attraction and repulsion was intended merely to establish the possibility of Newtonian physics (the laws of motion and the concept of universal gravitation), Helmholtz declared that his discussion of the forces of attraction and repulsion associated with matter demonstrated the actual conformity of nature to Newtonian central force laws, which were 'the necessary conceptual form for understanding nature'. By appealing to Kantian argument, Helmholtz acknowledged that his claim that Newtonian central force laws provided the only possible explanation of nature could not be established empirically, but demanded independent justification; his reference to metaphysical argument was meant to justify this claim and the physical actuality of the central force laws.

Helmholtz derived a general form of the principle of the conservation of living force, arguing that for the motion of a body acted on by a central force emanating from a fixed centre of force, the change of living force was measured by the change in a quantity he termed the 'tensional force'. The tensional force was the product of the intensity of the central force and the distance between the body and the force centre. The principle of the conservation of force – a principle that Helmholtz asserted was dependent on the supposition that the motions of the bodies were determined by laws of central forces – expressed the constancy of the sum of living force and tensional force.

Helmholtz's terms 'living force' and 'tensional force' correspond to 'kinetic energy' and 'potential energy'; and his principle of the 'conservation of force' provided a mathematical statement of the conservation of energy. Helmholtz was careful to specify his dual use of the term 'force' in his mathematical argument, to denote Newtonian central forces of attraction and repulsion as well as the energy quantities of living force and tensional force; and he was later quick to point out that 'living force' and 'tensional force' were synonymous with the 'energy' terms introduced in the 1850s.

Helmholtz applied his conservation principle to a wide variety of physical phenomena. Discussing the apparent loss of living force in inelastic collisions, he argued that the living force was consumed not only in the deformation of inelastic bodies (as Johann Bernoulli had suggested) – a process, Helmholtz explained, that would lead to the increase of tensional force – but also in the generation of heat. In a major advance on earlier discussion of inelastic collisions, mechanical 'losses' of energy were therefore explained by the conversion of mechanical energy into heat. Helmholtz went on to apply the conservation principle to thermal and electrical phenomena, drawing on Joule's early papers. He rejected the caloric theory of heat in favour of an explanation of heat transmission based on Ampère's theory that heat was propagated by means of a wave motion. In Helmholtz's theory, heat was explained in terms of matter in motion; thermal and mechanical phenomena were explicitly connected, forming part of a network of conversion processes subject to the law of the conservation of energy and subsumed under the mechanical theory of nature. Helmholtz's work on the conservation of energy was important not only in providing a mathematical formulation of the energy principle, but also in stressing the unifying role of the energy concept in relation to an ontology of matter in motion and the programme of mechanical explanation.

CHAPTER III

Energy Physics and Mechanical Explanation

In June 1847 William Thomson, later Lord Kelvin (1824–1907), met Joule at the Oxford meeting of the British Association for the Advancement of Science, and the encounter led Thomson to study Joule's papers on the mutual convertibility of heat and mechanical work. At the Oxford meeting Joule had read a paper describing his measurement of the temperature change in a fluid agitated by a paddle wheel that was turned by a descending hanging weight; he claimed to have determined the quantitative equivalence between the heat generated by the paddle wheel and the mechanical work required to generate that heat. Thomson found Joule's conclusions astonishing; and he reported Joule's work to his brother James Thomson (1822–92), who confessed that Joule's 'views have a slight tendency to unsettle one's mind'. The Thomsons' sense of intellectual disorientation arose from their belief, derived from the work of Sadi Carnot (1796–1832), that heat was conserved in the generation of mechanical work by heat engines. This theory seemed to contradict Joule's claim that heat must be consumed in the generation of work. The unravelling of the apparent contradiction between the theories of Carnot and Joule was to lead to the formulation of the science that in 1854 William Thomson was to term 'thermo-dynamics', the theory of the mechanical action of heat.

Carnot and the Motive Power of Heat

Sadi Carnot's *Réflexions sur la puissance motrice du feu* [*Reflections on the motive power of fire*] (1824) was written at a time when power technology and steam engines were the subject of intense interest among French engineers and physicists. Though Carnot's initial

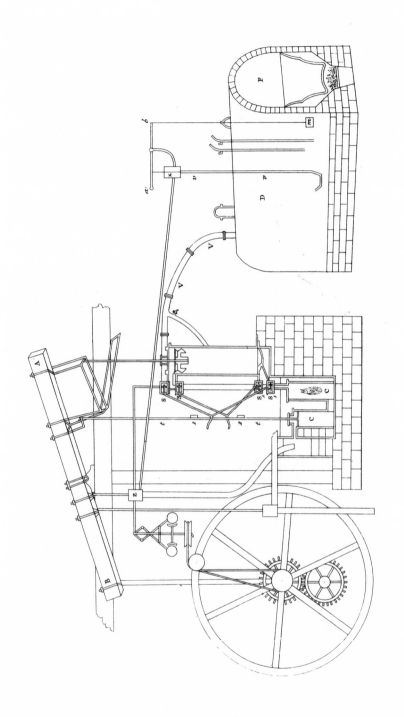

Fig. 3.1. Diagram of a steam engine designed by James Watt, as reproduced in Biot's *Traité de physique*. Sadi Carnot's theory of heat engines was shaped more by work on power engineering than by his indebtedness to contemporary physics, and he emphasised the importance of Watt's innovations. Steam engines were arousing strong interest in France, and several French physicists discussed the theory involved, including Biot in his important text. Carnot pointed out the social, technological, and scientific interest of steam engines, offering his treatise as a contribution to their theoretical understanding and practical improvement.

The importance of steam power was being belatedly recognised in France, and there was considerable interest in Arthur Woolf's double-cylinder high-pressure engines (which were briefly discussed by Biot); these, as Carnot observed, offered significant advantages in fuel economy (especially important in France, where good coal was in short supply). Carnot's explanation of the cycle of operations of an ideal heat engine was indebted to contemporary discussions of the high-pressure engine. Watt had suggested that the expansion stroke in a steam engine should continue after the cutoff of the steam supply, because the expansion of steam continued to drive the piston. Although Watt made little use of his 'expansive principle', the technique was of special importance in the high-pressure engines introduced into France after 1815, and the issue was discussed by physicists and engineers, including Carnot. Watt's expansive principle led Carnot to incorporate a phase of adiabatic expansion, following the phase of isothermal expansion, into the 'Carnot cycle' of operations for the ideal heat engine. *Source:* Biot, *Traité de physique*, 4:pl. VI, fig. 73.

problems and some of his arguments can be seen to have been shaped by the work of contemporary power engineers, the theoretical model he used in attempting a general analysis of the principles of heat engines was the work of his father, Lazare Carnot (1753–1823), on the theory of machines. Alluding to Lazare Carnot's attempt to elaborate a general theory to establish the maximum efficiency of machines, Sadi Carnot emphasised that a similar theory was needed for heat engines, one that would apply not only to steam engines but to all heat engines, whatever the working substance of the engine, and one that would be analogous to the general theory of machines.

Carnot's argument was structured by his adoption of the caloric theory of heat and his assumption that caloric (heat) was conserved in the production of mechanical work by heat engines. He observed that the fundamental feature of the steam engine was that steam alone would not generate work; there had to be a temperature difference for the engine to operate. Heat, or caloric, was not consumed in the operation of the engine: A quantity of caloric was supplied from the furnace to the boiler to produce the expansion of steam in the cylinder, and this same quantity of caloric was absorbed when the steam passed to the condenser. The steam merely had the role of transporting caloric between the boiler and the condenser. The crucial factor in the production of mechanical work by a heat engine was the temperature difference in the engine, and he envisaged the 'motive power' of heat as arising from the flow of caloric. He compared the fall of water in a water-powered engine to the 'fall' of caloric in a heat engine, a hydrodynamic analogy that suggests his indebtedness to the work of his father and other engineering theorists who had discussed the operation of engines driven by the pressure of water. Carnot supposed that just as the quantity of water was conserved in a water-powered engine, caloric was conserved in a heat engine. The motive power of heat was a function only of the temperature difference in the heat engine, and was independent of the working substance; any substance whose temperature could be changed by successive expansion and contraction in a heat engine could be employed.

In establishing this general principle of heat engines, Carnot described a cycle of operations carried out on an ideal engine consisting of a cylinder and piston, a working substance (atmospheric air), and two heat reservoirs maintained at different temperatures, the motive power of heat being produced by the 'fall' of caloric between the two reservoirs. In a cycle of processes the gas was successively expanded and compressed, and with each cycle caloric was transferred from the hot to the cold reservoir, generating mechanical work. The 'Carnot cycle' was reversible: By consuming the same quantity of mechanical work as was previously generated, the engine would return an equal quantity of caloric from the cold to the hot reservoir.

Carnot's description of the sequence of processes constituting the Carnot cycle differs significantly from the version published by Emile Clapeyron (1799–1864) in 1834. Clapeyron produced a mathematical reformulation of Carnot's argument that brought the Carnot cycle to the attention of physicists and engineers. He provided a graphical pressure–volume representation of the Carnot cycle (a form that soon became familiar in physics texts), and he expressed Carnot's principle – that the production of work was dependent only on the temperature difference in a heat engine – as a mathematical equation. Clapeyron also simplified the sequence of expansions and compressions in the Carnot cycle. In Clapeyron's version (though not in Carnot's), the engine and gas returned to their initial states at the end of each cycle, and heat (caloric) was explicitly conserved. Although Carnot himself ultimately abandoned the caloric theory of heat, it was Clapeyron's version of the Carnot cycle, depending explicitly on the conservation of caloric, that became the source of Carnot's ideas for Thomson and Clausius. Moreover, Carnot's discussion of the thermal properties of gases, in establishing that a given fall of temperature produced more work from a gas at low temperatures than at high temperatures, was couched in terms of the 'fall' of caloric; it therefore provided support for the caloric theory of gases and heat and reinforced the status of the caloric theory.

Thomson and the Problems of Thermodynamics

In his earliest writings on thermodynamics, published in 1848–9, William Thomson followed Carnot in accepting that heat was conserved in the generation of mechanical work by a heat engine. Although he considered this axiom of Carnot's theory to be 'still the most probable basis for an investigation of the motive power of heat', Thomson was troubled by the apparent conflict between Carnot's assertion of the conservation of caloric (heat) in the generation of mechanical work in heat engines and Joule's claim that whenever work was generated in a heat engine, a quantity of heat proportional to the work produced must be consumed.

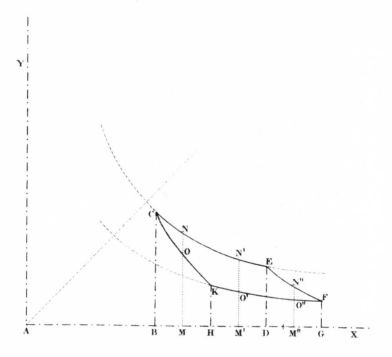

Fig. 3.2. Graphic representation of the Carnot cycle on a pressure–volume diagram, by Clapeyron (1834). An 'indicator diagram' had been used by Watt, but was kept a trade secret and was not used by Carnot himself. Here the ordinate represents pressure; the abscissa, volume. The gas is enclosed in a cylinder with a piston and is brought into contact with two heat reservoirs at temperatures T and t, where T is higher than t. The initial state of the gas is at C; its volume is represented by AB and its pressure by BC. The gas at temperature T is allowed to expand isothermally to E, and during the isothermal expansion a quantity of heat Q enters the gas from a heat reservoir at temperature T. The gas is expanded adiabatically to F, and during this expansion its temperature falls to temperature t. The gas is then compressed isothermally to K, a state defined by the condition that the same quantity of heat Q leaves the gas during its isothermal compression, and the heat flows to a heat reservoir at temperature t. The cycle is closed by the final adiabatic compression to C, during which the gas will return to its initial temperature T. Work is generated when heat (caloric) passes from the heat reservoir at temperature T to the reservoir at the lower temperature t; in Clapeyron's version of the Carnot cycle, heat is explicitly conserved in generating work.

The term 'isothermal' was standard usage in physical geography to denote equal temperatures. 'Adiabatic' was first used in 1858 by Rankine,

Thomson observed that 'the conversion of heat (or *caloric*) into mechanical effect [work] is probably impossible, certainly undiscovered', though he admitted that Joule's experiments seemed 'to indicate an actual conversion of mechanical effect into caloric'. While remaining publicly committed to Carnot's theory, Thomson recognised that the hypothesis of the conservation of heat, which he considered to be Carnot's 'fundamental axiom', was called into question by Joule's experiments.

Thomson's commitment to Carnot's viewpoint was threatened, however, by his caution over the caloric theory of heat itself. Thomson perceived that Carnot's axiom of the conservation, or indestructibility, of heat was grounded on the supposition of the caloric theory of heat as an imponderable substance. Thomson's work on Fourier's mathematical theory of the transmission of heat and its application to electrostatics had led him to emphasise the gap between mathematical and physical representations of nature; he was cautious about accepting Carnot's assumption that heat was an imponderable substance, and his caution led him to question Carnot's hypothesis that heat was conserved in the generation of mechanical work in heat engines.

Thomson also perceived a fundamental difficulty facing Joule's theory of the conversion of heat into work, a difficulty raised by irreversible phenomena such as the conduction of heat. Thomson pointed out that no mechanical effect was observed when heat was conducted through a solid: 'When "thermal agency" is thus spent in conducting heat through a solid, what becomes of the mechanical effect which it might produce? Nothing can be lost in the operations of nature – no energy can be destroyed. What effect then is produced in place of the mechanical effect which is lost?' This important query contains the first use of the term 'energy' as a general and fundamental physical concept (as distinct from Thomas Young's use of it, certainly known to Thomson, to denote living force). The query expresses Thomson's view that there was a difficulty in reconciling Joule's theory of the equivalence of heat and mechanical work, which denied that energy could be destroyed, with the phenomenon of thermal conduction, in which

---

Fig. 3.2 (*cont.*)
to denote the rise in temperature that occurs when a gas is compressed under conditions of thermal insulation. *Source:* Emile Clapeyron, 'Mémoire sur la puissance motrice de la feu,' *Journal école polytechnique* 14, cahier 23 (1834), 190, fig. 1.

heat was dissipated in conduction through a solid. Joule's theory seemed to imply that the heat expended in conduction should have been available to produce mechanical work.

Thomson's perception of the conceptual tangles besetting the emergent science of thermodynamics was profound. He did not merely delineate the contradiction between Carnot's axiom of the conservation of caloric in the generation of work in heat engines and Joule's claim that heat must be consumed in the generation of work, but queried the status of Carnot's assumption of the caloric theory itself. Moreover, he raised a fundamental difficulty that apparently threatened Joule's theory of the equivalence of heat and work: the explanation of irreversible thermal processes. Thomson was delineating problems rather than providing solutions, and in appealing to experimental evidence to resolve the apparent contradiction between the theories of Joule and Carnot, he hoped for a partial resolution of the contradictions he had formulated. But the experimental evidence provided no clear answer. James Thomson had used Carnot's theory to predict that the freezing point of water would be lowered by the application of pressure, a claim verified experimentally by William Thomson; yet Carnot's theory was apparently contradicted by Joule's experimental work.

Clausius and the Laws of Thermodynamics

The problems of constructing the science of thermodynamics were posed in a more limited and straightforward manner, and hence in a form more amenable to solution, by Rudolf Clausius (1822–88). In an important paper, 'Ueber die bewegende Kraft der Wärme' ['On the moving force of heat'] (1850), Clausius maintained that Joule's careful experiments had proved the equivalence between heat and mechanical work, and clearly contradicted Carnot's assertion that a loss of heat never occurs in the production of work by heat engines. Clausius declared that the production of work resulted not merely from a change in the distribution of heat, but also from the consumption of heat; and he argued that heat could be produced by the expenditure of mechanical work. Clausius concluded that there was no need to make a straightforward choice between the theories of Carnot and Joule, because the 'fundamental principle' of Carnot's theory was that heat would pass from a warm body to a colder one whenever work was done in a cyclic process. He declared that this principle could be retained even if Carnot's additional assertion, that the quantity of heat was undiminished in the process, was abandoned. Carnot's assumption that no heat was lost in a cyclic process was a subsidiary principle: 'It is quite possible,' wrote Clausius, 'that in the production of work . . . a certain portion of heat may be consumed, and a further portion transmitted from a warm body to a cold one: and both portions may stand in a certain definite relation to the quantity of work produced'. In this modified form, detached from the assumption of the conservation of heat, Carnot's fundamental principle would be compatible with Joule's theory that whenever work was produced by heat, a quantity of heat proportional to the work generated would be consumed. The conceptual basis of thermodynamics rested therefore not on a choice between, but on the reconciliation of, the theories of Joule and Carnot.

Clausius therefore stated two fundamental principles, the equivalence of heat and work, and the principle explaining the generation of work from heat in a cyclic process, according to which, in falling between two temperature levels, part of the heat is converted into work while the rest descends to the lower temperature. This second principle, Clausius noted, was the 'really essential' part of Carnot's theory, and he stated it as an empirical generalisation, justifying his assertion with the argument that heat always shows a tendency to equalise temperature differences and therefore to pass from hotter to colder bodies. These two principles as stated by Clausius became known as the two laws of thermodynamics.

Clausius drew a second important conclusion from Joule's demonstration of the equivalence between heat and mechanical work, arguing (as Joule had also done) that the equivalence of heat and work supported the hypothesis that heat consisted in a motion of the particles constituting bodies. Clausius maintained that the living force (kinetic energy) of the motion of these particles could be converted into mechanical work. He did not, however, base his formulation of the laws of thermodynamics on his 'mechanical' theory of heat. He noted that, although he took the view that the particles of a body were in motion and that heat was the measure of their living force, he did not intend to consider this matter further, other than to affirm his belief that the equivalence of work and heat implied this conception of the nature of heat. The hypothesis that heat consisted in the motion of the particles of bodies made the equivalence of heat and mechanical work conceptually intelligible, providing a mechanical basis for the relation between heat and work. Writing in 1857, Clausius declared that he had not emphasised his mechanistic ontology because he wished 'to separate the conclusions which are deducible from certain general principles from those which presuppose a particular kind of motion'. Clausius

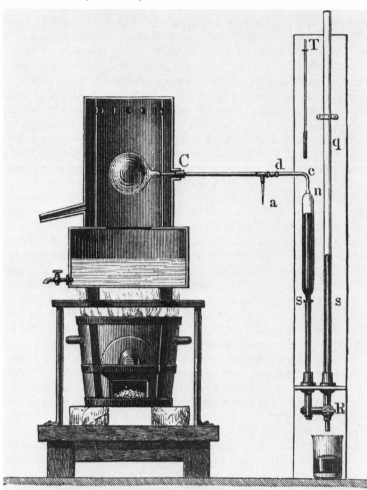

Fig. 3.3. An experiment to demonstrate the production of work from heat in heat engines, by Helmholtz. In an 1862 lecture on energy physics, Helmholtz illustrated the operation of heat engines by the generation of work from the expansion of gases. A glass globe filled with air was placed in a metal vessel in which it was heated by steam, and this globe was connected to a tube containing a liquid. On heating, the air in the globe expanded, generating work and displacing the liquid in the tube. This experiment illustrated the production of work by steam engines: The air in the globe would be replaced by water, which could be converted into steam by the application of heat, and the steam would drive a piston rather than a column of liquid.

therefore made an explicit distinction between the principles that he considered to be the fundamental laws of thermodynamics and specific assumptions about the nature of matter and heat. Though the formulation of the axioms or laws of thermodynamics was therefore independent of hypotheses about the nature of heat, nevertheless the equivalence of work and heat was supported by the mechanical theory of heat. The relationship between the laws of thermodynamics and the mechanistic ontology adopted by nineteenth-century physicists was to shape the conceptual development of thermodynamics.

Thomson: The Conservation and Dissipation of Energy

Clausius's achievement was to perceive that there were two independent laws that would form the basis of the science of thermodynamics. In his major paper 'On the dynamical theory of heat' (1851), William Thomson remarked that he, too, had realised that Carnot's theory could be modified so as to be consistent with Joule's theory of the equivalence of heat and work, though he acknowledged Clausius's priority. Thomson's own formulation of the laws of thermodynamics, influenced by Clausius's arguments on the theory of heat, was also shaped by his awareness of the work of W. J. Macquorn Rankine (1820–72). In 1850 Rankine had proposed a hypothesis of 'molecular vortices', demonstrating the equivalence of heat and work by equating the work produced by the rotation of the elastic atmospheres surrounding particles of matter with the heat produced; and in the following year he

---

Fig. 3.3 *(cont.)*

Helmholtz used a simplified form of an apparatus employed by Henri Victor Regnault (1810–78) in 1847 to measure the effect of heat on gases. Regnault had been concerned with exact measurement of the effect of heat on the expansion of different gases, and the effects of heating on air at different pressures. Regnault's work formed part of his exhaustive redetermination of the experimental data relevant to the study of gases and the theory and practice of steam engines. This work, though devoid of theoretical discussion, epitomised the new standard of precise and systematic experimentation in physics; students (including William Thomson) came to his laboratory from all over Europe to study his experimental techniques. In drawing upon Regnault's celebrated experimental work, Helmholtz was emphasising the empirical basis of energy physics. *Source*: Hermann von Helmholtz, *Populäre wissenschaftliche Vorträge*, 2 vols. (Brunswick, 1871), 2:157, fig. 18.

developed the argument in an attempt to include Clausius's second axiom (the modified Carnot principle). Most significantly for Thomson, in his 1850 paper Rankine had discussed the non-scalding property of steam escaping through an orifice from a high-pressure boiler, explaining it by the absence of liquefaction and concluding that heat must be supplied from an external source to prevent the liquefaction of the steam. Thomson maintained that heat was acquired by friction as the steam issued through the orifice, and concluded that this finding provided evidence in favour of Joule's theory that heat was evolved by the friction of fluids in motion. By late 1850, therefore, Thomson's perception of the state of thermodynamic theory had altered, and this alteration led to his acceptance of Joule's theory of the mutual convertibility of heat and work – and hence of Joule's claim that heat was consumed in the generation of work by heat engines.

In the draft of 'On the dynamical theory of heat', written in February and March 1851, Thomson set out the scope of the conceptual problems facing the science of thermodynamics, displaying a breadth of physical world view going beyond Clausius's more limited perception of the problem. The key issue for Thomson, as it would later be for Clausius also, was the problem of irreversibility: 'The difficulty which weighed principally with me in not accepting the theory so ably supported by Mr Joule was that the mechanical effect stated in Carnot's theory to be *absolutely lost* by conduction, is not accounted for in the dynamical theory otherwise than by asserting *it is not lost*'. In accepting Joule's doctrine of the equivalence and mutual convertibility of heat and work, Thomson affirmed, as the title of his paper made explicit, that the 'dynamical' or mechanical theory that heat consisted in the motion of the particles of bodies was the physical basis of Joule's theory. In the published paper Thomson observed that Joule had drawn the implication of the mutual convertibility of heat and mechanical work: that 'heat is not a substance but a state of motion'. The 'dynamical theory of heat' thus asserted that the work which might have been produced from the heat expended in conduction through a solid body was not lost, because the heat was transformed into the energy of motion of the unobservable particles of the body. Though the heat was unrecoverable, it was transformed and dissipated, not destroyed. Although Thomson avoided enunciating a theory of molecular processes to provide a mechanical model for the conversion of heat into the energy of the particulate motions of bodies (as Rankine had done), he nevertheless appealed to the dynamical theory of heat (as Clausius had done) as providing a justification for the laws of thermodynamics whose formulation was independent of specific hypotheses about the nature of matter and heat.

Thomson stressed that the phenomenon of irreversibility demonstrated the directionality of heat flow. His formulation of the modified form of Carnot's theory expressed the dissipation of heat from hotter to colder bodies. The dissipation of heat in processes that did not fulfil the criterion of perfect reversibility, in the steam engine or in the extreme case of thermal conduction, was thus envisaged as a fundamental feature of the material world. Heat was not conserved in heat engines, as Carnot had supposed, but was dissipated and also converted into work; for Thomson this doctrine expressed the essential meaning of Clausius's modified version of Carnot's theory, stating the directionality of heat flow from hotter to colder bodies. According to Clausius's modification of Carnot's theory, heat flowed from the hot boiler to the cold reservoir in the steam engine, but not all the heat was converted into mechanical work. According to Thomson, some of the heat was dissipated, remaining unrecoverable for the performance of work, though the dissipated heat was not destroyed.

For Thomson the two laws of thermodynamics expressed the indestructibility and the dissipation of energy. The two laws are compatible because dissipated energy is not destroyed, merely unrecoverable. Thomson's theory of the dissipation of energy merely declares, however, that the energy is wasted but not destroyed; Thomson did not demonstrate that the energy is not annihilated. As Maxwell put the problem to Thomson in 1855: 'Do you profess to account for what becomes of the *vis viva* [kinetic energy] of heat when it passes thro' a conductor from hot to cold?' For Thomson the dynamical theory of heat provided an indication of the means by which heat could be dissipated in conduction, and a physical model by which the axiom of the indestructibility of energy could be applied to irreversible processes. But for Thomson the laws of the indestructibility and dissipation of energy were also sanctioned by an appeal to theological argument. In the draft of his paper he argued that the immutability of energy was to be seen in terms of God's relation to nature: Energy was an immutable natural agent that could not be created or destroyed except by an act of divine power. Nevertheless, as he observed in the draft, 'everything in the material world is progressive', and the dissipation of energy expressed the directional, developmental character of the physical universe. Though theological arguments were not employed in the published paper, Thomson declared in a paper of

1852 that, 'as it is most certain that Creative Power alone can either call into existence or annihilate mechanical energy', dissipated energy cannot be annihilated, only transformed. The creation of energy and the restoration of energy dissipated in irreversible processes are thus acts which can only be performed by divine agency.

The Emergence of Energy Physics

In an address to the British Association in 1854, Thomson declared that Joule's discovery of the conversion of heat into work had 'led to the greatest reform that physical science had experienced since the days of Newton', the development of energy physics. In his introductory lecture at Glasgow in 1846, Thomson had argued that physics was to be based on the laws of dynamics, physics being the science of force. By 1851 energy had become, in his view, the primary concept on which physics was to be based, and his generalised use of the concept to apply to all phenomena of physics expressed this primacy. The fundamental status of energy derived from its immutability and convertibility, and from its unifying role in linking all physical phenomena within a web of energy transformations. The relationship between the indestructibility and the dissipation of energy broadened the application of the energy concept to all physical processes. The shift of emphasis from force to energy led Thomson to seek to establish the status of the energy concept within mechanics.

Thomson suggested that energy could be divided into two classes, which he termed 'statical' and 'dynamical'. Weights at a height, an electrified body, a quantity of fuel – all contained stores of 'statical' energy. Masses of matter in motion, a volume of space through which undulations of light or radiant heat are passing, and a body having thermal motions among its particles contained stores of 'dynamical' energy. Electrical, optical, and thermal phenomena were all linked by the concept of energy, with the implication that all forms of energy were forms of 'mechanical energy', and that all the phenomena of nature, not merely those falling under the traditional framework of mechanical problems, could be subsumed within a theory of mechanical explanation based on energy.

Thomson's science of physics based on the primacy of the energy concept was further developed by Rankine in papers published between 1852 and 1855. Rankine declared that the term 'energy' could be applied to 'ordinary motion and mechanical power [work], chemical action, heat, light, electricity, magnetism, and all other powers, known or unknown, which are convertible or commensurable with these'; he emphasised the universality and unifying role of the energy concept, which followed from the establishment of the law 'that all the different kinds of physical energy in the universe are mutually convertible'. The law of the transformability of energy applied to 'the objects of experimental physics'. Stressing the fundamental status of energy as a primary agent in nature, Rankine defined the framework of the science of 'physics', including mechanics, heat, light, and electricity, in terms of the universal energy concept and the conservation of energy in nature. Rankine replaced Thomson's classification of energy as 'statical' or 'dynamical' by the terms 'potential or latent' and 'actual or sensible' energy (although his use of these terms differed slightly from Thomson's classification), drawing on the familiar philosophical distinction between potential and actual existence; in Thomson and Tait's *Treatise on natural philosophy* (1867), these terms were replaced by 'potential' and 'kinetic' energy. Rankine observed that 'actual' energy could cause a substance to change its state and could thereby lead to the disappearance of actual energy and its replacement by potential energy; if this process were reversed, then as the potential energy disappeared, actual energy would be reproduced. Rankine then stated the fundamental law of the transformation of energy: 'The law of the conservation of energy is already known, viz. that the sum of the actual [kinetic] and potential energies in the universe is unchangeable'.

The theory of energy, or 'energetics', as Rankine described it in 1855, was developed as a framework of axioms free from the uncertainty of hypotheses about the nature of matter. Regarding his use of a molecular model in his early papers as a preliminary step useful in reducing the phenomena to simplicity and order, Rankine stressed the value of the energy concept in forming the basis of a systematic, axiomatic, and nonhypothetical physics. Energy was the common characteristic of the various states of matter to which the diverse phenomena of physics belonged, and the general laws of energy were therefore applicable to all branches of physics. A framework of axioms based on energy, the general principles of the 'science of energetics', would form the basis of physics. Thomson and Rankine envisaged a general theory of energy physics divorced from hypotheses about the nature of matter, a theory that defined the conceptual framework of physics and that would use the unifying role of energy and the law of the conservation of energy to redefine the programme of the mechanical explanation of phenomena.

Conversion and Conservation: Force and Energy Concepts

Rankine's statement of the 'law of the conservation of energy' in 1853 provided a synthesis of theories of conservation in physics. The law of the conservation of *vis viva* (living force) had been an established principle of mechanics for a century, its usage clearly distinct from the meaning of 'force' as defined by Newton's laws of motion; and Helmholtz had enunciated his principle of the conservation of force as a generalisation of the principle of the conservation of living force. The doctrine of the convertibility of natural powers or indestructibility of forces had sought to formulate a concept of the balance of natural agents, without a precise definition of the kind of quantitative equivalence envisaged. Joule's experimental demonstration of the equivalence of heat and mechanical work had provided a quantitative measure of the relationship between natural powers; and Helmholtz's statement of the mathematical principles of the theory of the conservation of energy had highlighted the ambiguity of the notion of the indestructibility or constancy of force, and the ambiguity of the term 'force' as descriptive of conserved physical quantities. Thomson's use of 'energy' resolved these terminological and conceptual confusions, and his emphasis on different manifestations of energy as different forms of 'mechanical energy' stressed the unifying role of the energy concept within a theory of the mechanical explanation of phenomena.

Rankine's arguments provided a synthesis and conceptual explication of the principles of energy theory, and Helmholtz underscored the conceptual and terminological implications of the new usage in responding to Rankine's introduction of the expression 'the conservation of energy'. Although the expression 'conservation of force' remained an appropriate designation of the indestructibility and transformability of natural agents, referring to the 'conservation of energy' gave a precise description of the quantities conserved. Within the framework of a mathematical theory of the conservation of energy, it became important to expurgate the doctrine of the conversion of forces from physical theory, distinguishing the energy principle from the concept of force as defined by Newton's laws of motion.

The reactions of Rankine and Maxwell to Faraday's essay 'On the conservation of force' (1857) illustrate this development. For Faraday the 'conservation of force' meant the transformability and indestructibility of natural powers. But in arguing that to construe gravitation as an action-at-a-distance phenomenon (in terms of the inverse-square force law of gravity) implies the creation and annihilation of force as two gravitationally acting bodies approach and recede from one another, and hence ignores the 'principle of the conservation of force', he exposed the ambiguities of term and concept in the expression 'conservation of force'. In response, Rankine urged that the phrase 'conservation of energy', about which there was no ambiguity, should be employed in future. He made it clear that the concept of force as employed in mechanics and in the theory of gravity was not conserved, whereas energy was a conserved quantity. Whereas force could be defined as the tendency of a body to change its place, energy was a conserved quantity measured by the product of force and the distance through which a force acted in bringing about a change. Force and energy were therefore physically distinct.

Maxwell raised similar objections in a letter to Faraday written in 1857, and Faraday's reply highlights the difference between his concept of the conservation of force and the law of the conservation of energy. By 'force' Faraday meant the 'source or sources of all possible actions of the particles or materials of the universe, these being often called the powers of nature'. He amplified this remark in an addendum to his paper in 1859, emphasising that he employed the term 'force' to mean 'the *cause* of a physical action', rather than to denote the tendency of a body to pass from one place to another. Those physicists who urged the primacy of the energy concept in the 1850s – Helmholtz, Thomson, Rankine, and Maxwell – wished to expunge the traditional but mathematically imprecise and ambiguous notion of the conversion or conservation of force from physical theory. The law of the conservation of energy was viewed as mathematically clear and precise, and as grounded on the experimental establishment by Joule of the equivalence of heat and work, which was itself seen as supporting the dynamical or mechanical theory of heat as the motion of the particles of bodies. Formulated as a principle of the conservation of mechanical energy and linked to the mechanical theory of heat, the law of the conservation of energy came to be regarded as fundamental to the mechanical explanation of physical reality.

The issues surrounding the interpretation of energy conservation as based on the mathematisation of physical quantities and exact, quantitative experimentation, which were important parameters of theory construction in nineteenth-century physics, can be seen in the controversy in the 1860s between John Tyndall (1820–93) and Peter Guthrie Tait (1831–1901) over the contributions of Julius Robert Mayer (1814–78) to the establishment of the

Fig. 3.4. Apparatus to illustrate the conversion of mechanical work into electrical energy, by Helmholtz. Turning the handle rotated a coil of copper wire between the poles of a magnet, generating electricity (which could then be used for the electrolytic decomposition of water, as shown in the apparatus). This device, the dynamo, was invented by Faraday in 1831 in his discovery of 'magneto-electric induction', the production of a current of electricity by a magnet. Faraday's discovery was part of his programme of exploring the connections between electric and magnetic forces, and Helmholtz used this example of the interconversion of natural agents to demonstrate the empirical basis of the law of energy conservation.

Faraday's discovery of the dynamo ultimately provided the basis of the electric power industry and led to the enormous technological and social

principle of energy conservation. Tyndall had defended Mayer's claim to be judged as an originator of the law of energy conservation, referring to Mayer's priority in calculating, in a paper published in 1842, the mechanical equivalent of heat from the heat evolved in the compression of a gas. Mayer had declared his intention to explicate the framework of a science of forces analogous to chemistry, the science of matter. Just as the causal relation between material causes and material effects expressed the indestructibility of substances in chemical processes, the proof of the indestructibility of forces such as motion, heat, and electricity, as quasi-substantial but nonmaterial entities, could be seen to depend on the relation between causes and effects. Mayer emphasised the equivalence of forces to one another in discussing the transformation of motion into heat, but he did not presuppose the mechanical theory of heat in calculating the mechanical equivalent of heat. The causal relation between forces implied their indestructibility and transformability, not their identity. In Mayer's view all forces had the same physical status, and their causal relation implied that they belonged to the category of force, not that heat could be defined in terms of motion.

Tyndall's defence of Mayer aroused Tait's xenophobia. Tait poured scorn on Mayer's ideas as subversive of the method of experimental science, and Joule claimed that Mayer had published an unsupported hypothesis, not an experimentally established quantitative law. In defence of his fellow German, Helmholtz reminded Tait of Mayer's importance as an innovator, though he later derided Mayer's 'metaphysically-formulated pseudo-proof' of the conservation principle. The mathematical formulation of the principle of energy conservation, the clarification of the energy concept, the experimental demonstration of the equivalence of heat and work, and the interpretation of energy as basic to the

---

Fig. 3.4 (cont.)

consequences of the science of electricity. The term 'dynamo' (not used by Helmholtz in this lecture of 1862) was first employed by a leading electrical technologist, Werner Siemens (1816–92), in 1867. Siemens, an associate of Helmholtz's, was an industrialist whose success in the electrical industry and concern with precision engineering, and whose belief in the technological utility of research in physics, led him to found the Berlin Physical–Technical Institute, which was completed in 1887. At the institute, of which Helmholtz was director, research into both fundamental physics and technology was pursued. Source: Helmholtz, Populäre wissenschaftliche Vorträge, 2:174, fig. 24.

programme of mechanical explanation made Mayer's ideas appear remote from the energy physics of the 1850s and 1860s.

Irreversibility: Clausius and Entropy

In his first formulation of the laws of thermodynamics in 1850, Clausius had stated the second law in terms of the directionality of heat flow, the tendency of heat to pass from a hotter to a colder body. He had presented the law as an empirical generalisation and had not been concerned with the conceptual implications of irreversibility. In subsequent papers, however, he attempted to enunciate a concept that would provide a measure of the direction of thermal transformations, and that would establish irreversibility as a fundamental feature of the natural world. In a paper on the second law of thermodynamics published in 1854, Clausius pointed out that this law expressed the direction of heat flow during the conversion of heat into work. He argued that the second law of thermodynamics could be represented as expressing a relation between the transformation of heat into work and the transformation of heat at a higher temperature into heat at a lower temperature. He suggested that these two transformations could be considered as equivalent in the sense that they could replace each other. The flow of heat from a hot to a cold body (accompanying the conversion of heat into mechanical work) could be counteracted by the conversion of work into heat, so that heat would flow from the colder to the warmer body. Clausius proposed to represent thermal transformations by the concept of the 'equivalence value' of a transformation: The equivalence value of the transformation of work into heat at a given temperature was measured by the heat produced, divided by the absolute temperature at which the transformation occurred; a measure of the equivalence value of the transmission of heat from a higher to a lower temperature was thereby provided.

The concept of the equivalence value of thermal transformations provided a clarification of the difference between reversible and irreversible processes. For a complete cycle of reversible processes the transformations would exactly cancel each other, so that the sum total of the equivalence values of the various thermal transformations constituting the reversible cycle would be zero. Although Rankine had proposed a similar concept, which he termed the 'thermodynamic function', he did not apply it to the analysis of irreversible processes. The distinctive feature of Clausius's argument was the application of the equivalence value of thermal transformations to the analysis of irreversible processes and hence to a formulation of the second law of thermodynamics. Assuming arbitrarily that the transformation of heat from a higher to a lower temperature should be assigned a positive equivalence value, the second law of thermodynamics was formulated as corresponding to a positive equivalence value of thermal transformations: 'The algebraical sum of all transformations occurring in a circular process can only be positive'.

In 1865 Clausius proposed the term 'entropy' (from the Greek word for transformation) in place of the term 'equivalence value'; hence the tendency of heat to pass from warmer to colder bodies was represented by an increase in entropy. He now emphasised the directional character of physical processes. Whereas the first law of thermodynamics expressed the conservation of energy in the universe, the second law expressed the dissipation of energy, denoted as a tendency to an increase in entropy in physical processes. Entropy denoted the directional character of physical processes and was introduced as a counterpart to energy because the two concepts had an analogous physical significance. Clausius thus stated the two laws of thermodynamics as 'The energy of the universe is constant', and 'The entropy of the universe tends to a maximum'.

Clausius adopted Thomson's term 'energy', and his new formulation of the two laws of thermodynamics stressed the significance of the energy concept, its conservation and dissipation. Realising the importance of Thomson's dissipation principle, Clausius employed the concept of entropy to emphasise irreversibility as a fundamental feature of nature. He continued to argue that the two laws of thermodynamics were axioms independent of any theory regarding the molecular constitution of bodies, but he attempted to provide a demonstration of the physical intelligibility of entropy by formulating a theory of underlying molecular motions. In a paper published in 1862 he introduced the idea of 'disgregation' as a measure of the arrangement of the molecules in a body, and he explained the physical meaning of entropy by the heat present in the body (which was measured by the molecular kinetic energy) and the disgregation (which provided a measure of the molecular configuration of the body). For Clausius, disgregation was a more fundamental concept than entropy, for which it provided a mechanical explanation. Nevertheless, he drew a distinction between his formulation of the laws of thermodynamics and his more special assumptions about unobservable molecular motions (which he believed made the laws of thermodynamics physically intelligible), pointing out that his attempt to provide a mechanical model was strictly supplementary to his prior establishment of the laws of thermodynamics as axioms independent of hypothetical molecular motions.

Clausius's attempt to define entropy in terms of molecular motions and arrangements led to considerable critical comment. Tait subjected Clausius to blasts of vituperation, defending the priority and conceptual sophistication of Thomson's thermodynamic thought and misunderstanding the entropy concept in the process, an error repeated in Maxwell's *Theory of heat* (1871), where Clausius's work was virtually ignored. Although Clausius's complaint led Maxwell to issue a corrected edition of his book, Maxwell remained critical of the concept of disgregation. He was concerned with stressing the gap between molecular science and the principles of thermodynamics. Even though Clausius had sought to separate his molecular hypotheses from his formulation of the laws of thermodynamics in terms of energy and entropy, Maxwell considered the disgregation concept inadmissible, its introduction merely confusing the conceptual structure of thermodynamics. The attempt to explain the laws of thermodynamics by molecular models was to him inappropriate. He concluded that 'we define thermodynamics, as I think we may now do, as the investigation of the dynamical and thermal properties of bodies, deduced entirely from what are called the First and Second laws of Thermodynamics, without any hypotheses as to the molecular constitution of bodies'.

Clausius's theory of the molecular processes underlying thermal phenomena, which he developed at length in his work on the theory of gases, demonstrates his belief that the laws of thermodynamics should be explained by a theory of molecular configuration. In contrast to Clausius, Maxwell argued that the second law of thermodynamics was an essentially statistical law that described the behaviour of an immense number of molecules, and could not be explained by a theory of individual molecular motions; he therefore sought to clarify the status of the second law in relation to his statistical theory of the motions of gas molecules.

Irreversibility and Cosmogeny

William Thomson consistently noted the importance of irreversibility in his statement of the second law of thermodynamics. He saw the dissipation of energy as implying a progressivist cosmogeny, an idea that harmonised with the biblical view of the transitory character of the universe. Energy was conserved, but irreversibly and progressively dissipated. He drew the implications for terrestrial history in an 1852 paper on the 'universal tendency in nature to the dissipation of mechanical energy', arguing that the law of the dissipation of energy and the theory of the progressive cooling of the earth implied that within a finite period of time the earth would be unfit for human habitation. The earth's heat was being irreversibly dissipated, a conclusion that Thomson believed was in conflict with Charles Lyell's geological theory, which emphasised the stability of the earth through a balance of geological processes. In the 1860s Thomson developed this theory so as to calculate the age of the earth, claiming that the time span of terrestrial history would be insufficient to allow Lyell's theory of gradual geological change and also Darwin's theory of evolution by natural selection.

There was a spate of interest in cosmological theory in the 1850s, and Thomson developed a theory of solar energy, arguing that meteors would spiral into the sun's atmosphere, finally falling into the body of the sun and producing the sun's heat. He maintained that all the energy transformations in the solar system could be traced to solar energy, and that the sun's heat was maintained by friction between the atmosphere of the sun and an ethereal vortex of evaporated meteors. Helmholtz also discussed cosmological questions in the 1850s, bringing energy considerations to bear on the 'nebular hypothesis' suggested by Kant and Laplace – the theory that the solar system originated from the condensation of gaseous matter that formed the sun and planets. Helmholtz argued that the sun was an incandescent molten mass originating from the collision of meteors, and that solar heat was produced by this process of collision. Thomson later adopted this version of the meteoric hypothesis, for he had come to realise that the sun's energy could not be maintained by meteors. In this modified version, as Thomson was quick to perceive, the meteoric theory implied a limit to the possible age of the solar system. Helmholtz, too, had noted the consequences of the dissipation principle; he observed that the transformation of mechanical energy into heat would have the result that all energy would finally pass into the form of heat, leading to the complete cessation of all natural processes: 'The universe from that time forward would be condemned to a state of eternal rest'. Clausius reformulated the cosmological consequence of the second law of thermodynamics, the 'heat death' of the universe, in terms of the entropy concept in 1867. He observed that, when the universe attained the condition at which its entropy was a maximum, no further change could take place, and 'the universe would be in a state of unchanging death'.

Some physicists sought to question this pessimistic conclusion. Helmholtz himself observed that it was possible that the universe had a boundary, and that although light and heat from the solar system were dissipated throughout space, such a boundary might permit the dissipated energy to be restored. A similar hypothesis had been suggested by Rankine, who maintained that the tendency in nature towards the conversion of all energy into the form of heat did not necessarily imply the cessation of all physical processes. It was likely that light and radiant heat were transmitted through the interstellar ether, and it was possible that the ether possessed a boundary. On reaching this boundary the radiant heat of the universe would be totally reflected and reconcentrated, and hence the physical universe would be able to reconcentrate its energy and renew its activity. The dissipation of energy would therefore be counteracted; energy would be restored and the universe would be reconstructed. Rankine speculated that there could be a steady-state situation, with the opposing processes of the dissipation and reconcentration of energy proceeding at the same time, so that the energy of the universe would be continuously depleted and restored.

It was, however, Thomson's views that shaped the argument of *The unseen universe,* published in 1875 by Tait and Balfour Stewart (1828–87). This work attempted to refute the criticisms that the doctrines of physics were materialistic and hostile to religion. Accepting Thomson's directionalist cosmogeny, the authors maintained that the physical dissipation of energy in the universe indicated that the physical or visible cosmos was transitory. They suggested, however, that dissipated energy would flow through an invisible ether to an invisible realm. Whereas the energy of the totality comprising the visible and invisible universes was indestructible and conserved, the energy of the visible universe was transitory and progressively dissipated. Whereas the physical universe would decay, the unseen universe would be eternal. In subsuming acts of divine providence under the law of the conservation of energy (which governed the transference of energy between the visible and the unseen universe), these apologetic arguments were controversial. Nevertheless, the book illustrates the cultural ramifications of nineteenth-century physics. Thomson had asserted that only divine agency could create or destroy energy, and Joule and Faraday had argued that divine providence was manifested in the wisdom and foresight by which cosmic order was established by the indestructibility of force. Appeal to theological arguments, not uncommon among British physicists, was a manifestation of the continued influence of the 'natural theology' tradition. Maxwell suggested that the identity of stellar and terrestrial spectra demonstrated the identity of molecules throughout the universe, and he concluded that molecules were like a 'manufactured article', their identity demonstrating their design by divine agency. These arguments illustrate the theological commitments that some physicists attempted to integrate into their physical conceptions of nature. The conservation and dissipation of energy, being concerned with the creation and end of the physical universe, provided an important focus for these arguments.

Energy Physics and 'Dynamical' Explanation

Thomson and Tait's *Treatise on natural philosophy* (1867) reconstructed analytical dynamics, emphasising the physical basis of Lagrange's generalised equations of motion and incorporating the law of the conservation of energy into the framework of analytical dynamics. These methods had a major impact on the 'dynamical' theories favoured by British physicists, which used explanations based on analytical dynamics as an alternative to the construction of specific mechanical models. Thomson and Tait had first discussed their project in 1861 and had soon agreed on an important point of terminology, the decision to employ 'dynamics' rather than simply 'mechanics'. This indicated the distinction between the mathematical science of abstract dynamics and the empirical study of machines, and also highlighted the importance of the idea of force. Rather than adopting a completely abstract approach, deriving generalised equations of motion that could then be employed as analytical theorems for solving problems, Thomson and Tait stressed a physical approach based on the assumption of Newton's laws of motion. They observed that the principle of the parallelogram of forces, the principle of virtual work, and d'Alembert's principle could all be derived from Newton's laws of motion (regarded as physical concepts established by experiment), rather than derived as analytical theorems.

Thomson and Tait considered the treatment of motion and the action of forces to be the fundamental problems of mathematical mechanics, corresponding to the division of the subject into kinematics (the study of motion in the abstract) and dynamics (the science concerned with the agency of forces). Dynamics, the study of forces, was itself subdivided into statics (the action of forces in maintaining states of rest) and kinetics (the action of forces in maintaining states of motion). The stress on force in Thomson and Tait's argument can be seen from their discussion of the status of statics as a branch of dynamics. They argued that statics should be derived from Newton's second law of motion and considered in terms of the action of force in maintaining dynamical equilibrium.

The basic mathematical axiom in their treatment of dynamics was a theorem discovered by Thomson, which related the variation of a system by impulsive forces to the kinetic energy of the system, enabling generalised equations of motion to be derived from the supposition of impulsive forces. Because an impulsive force acts in an infinitesimal time increment, he argued, the configuration and potential energy of the system would be unaltered by the action of an impulsive force, and only the kinetic energy would change. Thomson and Tait noted that the strength of this method was that it avoided the construction of a mechanical model of the system, while at the same time keeping the emphasis on physical reality by basing the mathematical argument on the action of impulsive forces. Maxwell expressed the issue with characteristic clarity in his *Treatise on electricity and magnetism* (1873), a work indebted to the dynamical methods introduced by Thomson and Tait. Maxwell pointed out that the hidden structure of a mechanical system could be represented in principle by an infinite number of possible mechanical models, whereas the formulation of a dynamical theory avoided consideration of the hidden mechanism by which the parts of the system were connected. In Maxwell's view, Thomson and Tait's formulation of abstract dynamics in terms of the theory of impulsive forces developed the Lagrangian generalised equations of motion in a way that stressed the physical basis of analytical dynamics and the physical significance of force.

In emphasising that the description of a specific mechanical model was to be avoided, Thomson and Tait gave the law of the conservation of energy a prominent place in their argument, stating that the energy of a material system, determined by the configuration and motion of the parts of the system, could be specified without reference to the hidden mechanism determining the configuration and motion of the system. The coordinates that specify this mechanism are ignored in the formulation of the equations of motion. They claimed that the law of the conservation of energy comprehended the whole of abstract dynamics, because the conditions of equilibrium and motion of a system could be derived from this law. Thomson and Tait also stressed the status of energy as a fundamental physical concept, declaring that 'energy is as real and as indestructible as matter'. Energy and matter were the fundamental constituents of nature, and the programme of mechanical explanation was to be based on the energy concept and its role within the formalism of abstract dynamics.

In his elementary treatise on mechanics, *Matter and motion* (1877), Maxwell summarised Thomson and Tait's view of the status of the energy conservation principle in its application to the unobservable particles of a material system. He observed that the energy of such a system was determined by the configuration and motion of the system, and that even though the mechanical structure of an electromagnetic system was unknown, the methods developed by Thomson and Tait did not require knowledge of the hidden mechanism. Because it treated an electromagnetic system in terms of the potential energy of configuration and the kinetic energy of motion, dynamics could be applied to electromagnetic phenomena without hypotheses about the hidden mechanical structure of the system; this was a universal programme of dynamical explanation grounded on the status of energy as a fundamental constituent of physical reality, and on the law of the conservation of energy as a unifying dynamical principle.

CHAPTER IV

Matter and Force:
Ether and Field Theories

The term 'magnetic field' was introduced by Faraday in 1845, and subsequently adopted by Thomson and Maxwell, whose usage clearly echoed Faraday's. Thomson first used the expression 'field of force' in a letter to Faraday in 1849, following their discussion of the nature of magnetism; and Maxwell first referred to a 'magnetic field' in a letter to Thomson in 1854, in the context of a discussion of Faraday's ideas. Maxwell gave the term 'field' its first clear definition, in consonance with previous usage, in his paper 'A dynamical theory of the electromagnetic field' (1865); there he stated, 'The theory I propose may therefore be called a theory of the *Electromagnetic Field,* because it has to do with the space in the neighbourhood of the electric or magnetic bodies'. The concept of a field was to be contrasted with an action-at-a-distance theory of electric action; that is, the mediation of forces by the agency of the contiguous elements of the field existing in the space between separated electrified bodies was to be distinguished from the action of forces operating directly between electrified bodies across finite distances of space.

The inclusive breadth of Maxwell's definition of the field makes it apparent that the physical status of the field was not defined uniquely. In a field theory the forces between bodies were *mediated* by some property of the ambient space or field. The field could be characterised as a field of force and represented in terms of the spatial distribution of forces; or alternatively the mediating property of the field could be characterised in terms of an intervening ether, envisaged either as a continuum or as constituted of discrete particles. In a field represented by a particulate ether, the ether particles acted only on neighbouring particles; the forces were exerted only between contiguous particles of the ether. Faraday

developed theories that fell into both categories: a theory that
supposed the primacy of lines of force in space, and a theory that
postulated the mediation of electric action by means of the
contiguous particles of a space-pervading substance, the 'dielectric'
medium. These formulations shaped the subsequent development
of the field concept, and Maxwell's definition embraced both these
categories of the physical status of the field.

An important feature of the field was the mathematical formalism adopted for its representation. Maxwell argued that the
mathematical language of partial differential equations was the
expression of the physical structure of the field. The equations of
the electromagnetic field involved quantities that were continuous
functions of their variables, and these equations expressed the
continuous propagation of force or energy between neighbouring
infinitesimal elements of space in the field. The physical structure
of the field was not defined uniquely: It could be represented as a
force plenum, a fluid continuum, or a particulate ether. Maxwell
observed that partial differential equations related electric and
magnetic variables at any given point to variables at adjacent points,
and noted that these equations could represent 'a theory of action
exerted between contiguous parts of a medium', providing a
mathematical formalism congruent with his representation of the
field by a mechanism of particles of matter in motion.

Faraday's Theories of the Field

Faraday's discovery of electromagnetic induction in 1831 raised the
question of the mode of propagation of electromagnetic forces,
that is, of their relationship to the material substance through
which they were propagated – the iron ring round which the wires
of the primary and secondary electric circuits were wrapped.
Faraday suggested that the passage of the electric current through
the primary circuit gave rise to a state of electric 'tension' in the
particles of the iron ring, an electrical condition of matter that he
termed the 'electro-tonic state'. The creation of the electro-tonic
state induced a current in the secondary circuit, and when the
current ceased to flow in the primary circuit the electro-tonic state
was dissipated. The transmission of electromagnetic forces was
therefore represented as taking place because the particles of the
iron ring were thrown into a state of electrical tension.

In his work on electrochemistry and electrostatic induction in
the 1830s, Faraday elaborated his concept of the electro-tonic
state, describing it as a polarisation of the molecules of matter, by

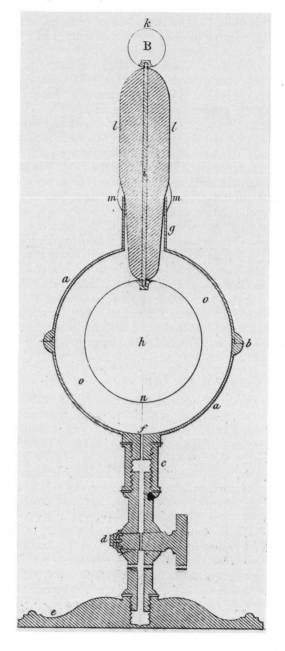

which he meant a disposition of force in which a molecule acquired opposite electrical powers on different parts, and he represented electrochemical decomposition as the transmission of forces by the particles of matter. In demonstrating that electrostatic induction took place along curved lines, he claimed that electrostatic induction was mediated by the transmission of force between the particles of a 'dielectric' medium surrounding the electrified bodies. Faraday showed experimentally that different substances had different capacities for the mediation of electrostatic forces. He argued that the particles of the dielectric transmitting electrical action were subjected to a state of electrical tension that led to the propagation of electrostatic forces. Faraday stressed the spatial distribution of electrical forces, a concept Oersted had also emphasised, and Ampère had speculated on the mediating role of an electrical ether in propagating electrical forces.

Faraday also employed the idea of lines of force, as a geometrical representation of the lines of polarised particles subjected to electric tension. But in the 1830s he used 'the term *line of inductive force* merely as a temporary conventional mode of expressing the

---

Fig. 4.1. Faraday's demonstration of specific inductive capacity (1837). Faraday sought to establish his theory of the polarisation of the particles of the 'dielectric' and the 'electro-tonic state' by demonstrating that substances possessed a specific inductive capacity. The apparatus consisted of two brass spheres *a* and *b*, one placed inside the other; the inner brass ball *b* was connected by a brass stem *i* to another brass ball *B*, the brass stem being surrounded by an insulating mass of shellac *l*. By electrifying the apparatus Faraday was able to measure the variations in charge corresponding to the introduction of different insulating substances in the space between the two brass spheres. He described in detail his experimental precautions to ensure accuracy and uniformity in the experimental tests, for the experiment required great delicacy in order to eliminate distorting effects.

The demonstration of the phenomenon of specific inductive capacity served to show that there was a relation between electric induction and the substances through which it was transmitted. Faraday concluded that electric induction was an action of the contiguous particles of the insulating dielectric, the substance through which electric forces acted. The experiment provided powerful support for his rejection of an action-at-a-distance theory of electricity, and for his emphasis on the mediating role of the dielectric medium in transmitting electrostatic forces. *Source*: Michael Faraday, *Experimental researches in electricity*, 3 vols. (London, 1839–55), 1:pl. VII, fig. 104.

direction of the power', affirming his commitment to the electro-tonic state and the transmission of force by 'an action of contiguous particles consisting in a species of polarity'. He did not represent the particles as being polarised as a result of mutual contact; the particles did not act by contact but by their associated polar forces. The distance between the neighbouring particles was not considered to be significant; though he denied that electrostatic action could be explained by the transmission of forces across 'sensible' distances, rejecting an explanation in terms of action at a distance, he also stated that the particles could act across a distance of half an inch. In response to criticism that he had contradicted himself, Faraday denied any contradiction in his argument and reaffirmed his idea of action between contiguous particles. Although Faraday's concept of the transmission of force between contiguous polarised particles provided no explanation of the mechanism by which the forces were propagated, he apparently envisaged electrical forces as manifested in the spaces surrounding the particles of electrified matter. Such conceptions were characteristic of late eighteenth-century electrical theory.

This criticism did, however, expose a lacuna in Faraday's theory, the absence of an explanation of the interaction of the forces between polarised particles. Faraday had stressed the concept of polarity rather than the discussion of the relationship between the particles and their associated forces. In facing this problem he was led to consider fundamental questions about the nature of matter. In the 1830s he had refrained from discussing the nature of matter; though he had indicated that electricity was associated with the particles of matter, he had focused his argument on electrical action and the polarity of particles, rather than on a theory of matter as such. His 'Speculation touching electric conduction and the nature of matter' (1844) formulated a theory of matter and the agency of force that aimed to provide a representation of the transmission of forces in space; in so doing Faraday renounced the atomic theory of matter and transformed his theory of the propagation of forces.

In the 'Speculation' Faraday pointed out that, according to the atomic theory, atoms were not considered to be in contact, and that, if action between contiguous particles was denied, then it would be necessary to ascribe a role to the spaces between the atoms to account for the communication of forces between particles. In his view, space could not have causal or dispositional properties analogous to a material substance (as he later remarked, 'mere space cannot act as matter acts'), and he therefore concluded that the theory of atoms and the void should be abandoned. He claimed that all knowledge of matter was limited to ideas of the system of 'forces or powers' associated with material substances, and he asserted that matter should not be considered as consisting of extended, impenetrable atoms surrounded by forces of attraction and repulsion; instead, matter should be envisaged as a plenum of 'powers' filling space: 'The substance consists of the powers'. This theory denied the impenetrability and indivisibility of atoms, supposing the 'mutual penetrability of matter'. This overcame the problem of explaining the mode of transmission of the forces between contiguous particles, for 'matter will be *continuous* throughout, and in considering a mass of it we have not to suppose a distinction between its atoms and any intervening space'. By virtue of its 'powers', the defining properties of matter, matter extended continuously through space, and interactions between 'particles' of matter were envisaged as interactions between 'centres of force' or arrangements of powers diffused through space.

Faraday indicated that this theory of matter resembled a theory proposed by R. J. Boscovich in the mid-eighteenth century; but Boscovich's ideas were quite different from Faraday's, because Boscovich did not define matter in terms of inherent powers but preserved the Newtonian dualism between force and matter, supposing that matter consisted of nonextended centres from which forces of attraction and repulsion operated. Faraday's theory of matter shows much closer similarities to arguments advanced by Joseph Priestley in the 1770s. Priestley had maintained that the defining characteristics of matter were extension and inherent powers of attraction and repulsion. He rejected the assumptions of Newtonian atomism – that impenetrability and solidity were essential properties of matter – and replaced the Newtonian dualism of atoms and forces by using force to define the essence of matter. Priestley had ambiguously presented his theory as similar to Boscovich's ideas, and Priestley's arguments, customarily designated as 'Boscovich's theory', were widely disseminated in texts between 1780 and 1830. Faraday's acknowledgment to Boscovich, though misleading, was conventional.

The theory of matter of the 'Speculation' initiated a radical reconstruction of Faraday's theory of the propagation of forces. In his 'Thoughts on ray-vibrations' (1846), Faraday stated that his theory of matter as a collocation of forces had led him to consider that the propagation of forces could be represented as vibrations in lines of force. The concept of lines of force presented matter as diffused through space and thus showed the disposition of material substances in space: 'The particle indeed is supposed to exist only by these forces, and where they are it is'. The lines of force that permeated space represented the interaction between material substances. In the 1830s Faraday had employed the concept of lines of force merely to represent the alignment of polarised particles, but his new theory of matter had the implication that the lines of force represented the structure of material substances and their interaction. This led him ultimately to abandon the polarisation of the particles of matter represented by the electro-tonic state, in favour of a theory of the primacy of lines of force to represent the propagation of action. As a result of a study of magnetic phenomena he began to use the term 'polarity' merely to represent the direction of the lines of force in the field of forces.

Faraday developed this theory of the primacy of lines of force in the course of a series of remarkable experimental investigations. In 1845, at William Thomson's suggestion, he renewed an attempt to detect the effect of magnetic forces on polarised light, observing the rotational effect of magnets on light. This discovery and his subsequent discovery of the effect of magnetism on the alignment of crystals led him to explain these phenomena by the direction of the lines of magnetic force. He explained the differences in the magnetic character of different substances by the propensity of lines of force to pass through these different substances – by the magnetic conductivity of lines of force in different substances. Faraday illustrated lines of force with the patterns assumed by iron filings sprinkled over magnets, and he summarised his new conceptual framework in a paper entitled 'The physical character of the lines of magnetic force' (1852). He declared that forces 'can only have relation to each other by *curved* lines of force through the surrounding space; and I cannot conceive curved lines of force without the conditions of a physical existence in that intermediate space'. The lines of force were the primary entities representing physical reality, rather than mere symbols that expressed the alignment of polarised particles. Faraday stated that he did not intend to 'confound space with matter'; his idea of matter filling space continuously by its forces did not suppose the identification of the structure of space with the disposition of the lines of force in space.

Faraday had introduced two distinct representations of the field, one using mediation by the contiguous particles of an ambient medium and one asserting the primacy of lines of force. Neither theory provided an explanation of the mechanism by which forces were propagated, though Faraday did suggest that the transmission of force 'may be a function of the aether'. In responding to

Faraday's theories of the field, Thomson and Maxwell attempted to develop physical theories of the propagation of forces by means of mechanical theories of the ether, and to render Faraday's concepts physically intelligible by reformulating them in terms of the programme of mechanical explanation. Maxwell stressed the importance of formulating a 'consistent representation' of the mechanism of the field, an expression he derived from a remark by Carl Friedrich Gauss (1777–1855) on the need to achieve a 'constructible representation' of the propagation of electric forces.

Thomson: Ether and Field

Thomson's first work on the theory of electricity in the 1840s had explored the mathematical analogy between thermal and electrical phenomena, drawing upon Fourier's theory of heat to develop a mathematical analysis of the flow of electrical force from a source corresponding to the flux of heat. On turning to a study of Faraday's theory of electrostatics, Thomson perceived that the mathematical analogy between thermal and electrostatic phenomena suggested a corresponding physical analogy. The physical model of the propagation of heat from particle to particle suggested an analogous propagation of electrical forces 'by the action of the contiguous particles of some intervening medium', as in Faraday's theory of the propagation of electrostatic forces by the contiguous particles of the dielectric medium. Thomson was, however, careful to distinguish between the expression of a mathematical analogy and the formulation of a physical hypothesis, pointing out that he had made no attempt to elaborate a hypothesis about the propagation of electrostatic action between the particles of the dielectric medium. Although the mathematical correspondence between thermal and electrical phenomena did not justify the supposition of a physical hypothesis for the propagation of electrostatic forces, the mathematical correspondence suggested that an appropriate model might well be based on a theory of action between contiguous particles. These remarks drew attention to the gap between mathematical and physical representation, a fundamental issue in Fourier's analytical theory of heat, which Thomson had employed to explore the mathematical and physical structure of the theory of electricity. Thomson's work also noted that Faraday had not provided an explanation of the mechanism of the molecular interactions between the contiguous particles of the dielectric medium.

Thomson attempted to meet this difficulty in his 'Mechanical

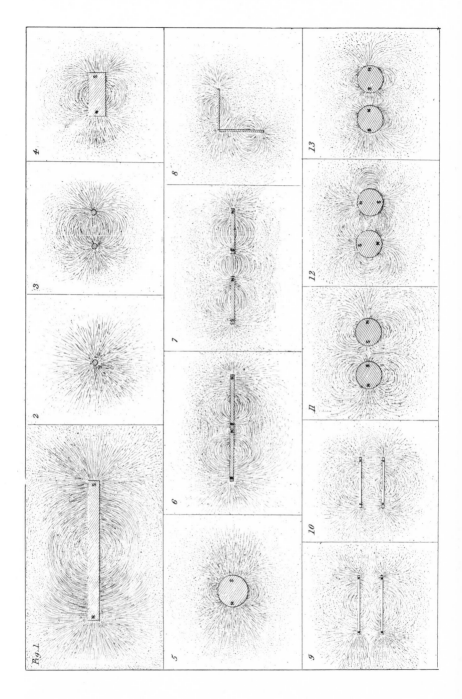

Fig. 4.2. Faraday's delineation of lines of magnetic force by iron filings (1852). These illustrations show the forms assumed by iron filings sprinkled over a magnet. Figure 1 shows the pattern assumed about a bar magnet; Figure 2, the pattern over a pole; and Figure 3, the pattern between contrary poles. Figure 4 shows the pattern with a short magnet; Figure 5, that with a magnetised steel disc. The other figures display the patterns when the magnetising power is varied. Faraday used these patterns to indicate the direction of the lines of magnetic force, in an experimental demonstration of his theory of magnetic power as represented by lines of magnetic force. As was customary in his descriptions of experimental arrangements, even in this simple demonstration, he gave a detailed account of the conditions of the experiment: the nature of the paper on which the iron filings should be sprinkled, the attention to the cleanness of the filings, and the sprinkling of the filings by a fine sieve. He declared that the use of iron filings made the lines of magnetic force visible to the eye.

Before 1852 Faraday had emphasised that the term 'line of force' represented merely the disposition of the forces, but he now argued that lines of force had a real physical existence. He appealed to his delineation of magnetic lines of force by iron filings in support of his contention that magnetic forces were related by curved lines of force. *Source:* Faraday, *Electricity.* 3:pl. III.

representation of electric, magnetic, and galvanic forces' (1847). This paper considered the propagation of electrical and magnetic forces in terms of the linear and rotational strain of an elastic solid, employing mathematical methods that Stokes had developed for the treatment of rotations and strains in continuous media. Faraday's discovery of the rotational effect of magnetism on light was consistent with the representation of magnetic force by the rotational strain of an elastic solid. The suggestion of a mechanical theory of the transmission of electrical and magnetic forces was intended, however, as Thomson informed Faraday in a letter in June 1847, as 'merely a sketch of the mathematical analogy' between electric and magnetic forces and the strain propagated through an elastic solid. Thomson indicated that he was not suggesting a physical theory of the propagation of electric and magnetic forces; though the mechanical model of the strain in continuous media had physical implications, the relationship between mechanical strain and electrical and magnetic phenomena remained physically unclear.

Thomson then turned to the mathematical representation of Faraday's theory of the magnetic field as a structure of lines of force. In a letter to Faraday of June 1849, Thomson made a sketch of lines of force in a magnetic field, which he termed a 'field of force', to illustrate the conductivity of substances to magnetic lines of force; and he developed this concept of the magnetic field in a paper called 'A mathematical theory of magnetism' (1851). He represented the magnetic field as the continuous distribution in space of an 'imaginary magnetic matter'. In positing that magnetic forces were exerted between portions of magnetic matter, Thomson did not suppose a physical model for the transmission of magnetic force between the particles of a magnetic substance. The 'magnetic matter' was not conceived as a material substance, but was envisaged as an embodiment of Faraday's theory of the primacy of lines of force, a plenum that represented the spatial distribution of the field of force, permitting the mathematical expression of lines of force in space. Thomson stressed the gap between mathematical and physical representation. His theory of the strain of an elastic solid and his representation of lines of magnetic force by a magnetic plenum were not intended as models of the physical constitution of the field, but were meant merely to provide physical illustrations or analogies for mathematical theories of the field.

In an important paper published in 1856, Thomson attempted to provide what he termed a 'dynamical illustration' of the physical field. His discussion of the molecular motions constituting the physical field exercised an important influence on subsequent theories of the field. Drawing on the concept of molecular motion he had adopted in his 'dynamical' theory of heat, and on Rankine's theory of heat as a vortical motion of ethereal atmospheres surrounding molecular nuclei, Thomson argued that the magneto-optic rotation discovered by Faraday could be explained as a vortical motion in the ether: 'The explanation of all phenomena of electro-magnetic attraction or repulsion, and of electro-magnetic induction, is to be looked for simply in the inertia and pressure of the matter of which the motions constitute heat'. The dynamical theory of heat provided the model for a physical theory of the field. Thomson did not, however, adopt the concept of 'molecular vortices' originally formulated by Rankine. As in his theory of heat, where Thomson had carefully avoided the specificity of Rankine's molecular model while committing himself to the general hypothesis that heat was a form of molecular motion, he deliberately left open the question of the physical structure of the ether whose motions constituted the field. He speculated that the ether could be a continuous fluid permeating the spaces between molecules of gross bodies or could be constituted of discrete molecules, or possibly that all matter was ultimately continuous, its apparent molecular structure produced by the vortical motions of a continuous ether.

In the late 1850s Thomson favoured the theory that the ether was to be envisaged as a fluid. His preoccupation with the problem of the material structure of the ether and the relationship between ether and matter can be seen from an unpublished manuscript of 1858, where he elaborated on his suggestion that all matter was continuous, the 'doctrine of the Universal Plenum'. He speculated that it might be possible to conceive matter in terms of 'motions or eddies in a fluid'. According to the theory of the ethereal plenum, the molecular structure and the solidity and impenetrability of bodies were the result of a vortical motion in a perfectly elastic ethereal continuum. He found no real physical possibility in these speculations, however, because vortices in a plenum did not seem to possess the property of indestructibility fundamental to any theory of the primordial constituents of matter.

Thomson's paper 'On vortex atoms' (1867) provided the solution to these speculations. He was stimulated by a paper of Helmholtz's, which showed that vortex filaments in a perfect fluid would be immune to destruction or dissipation, and would therefore satisfy Thomson's criterion of the immutability of primordial atoms, whose creation and destruction could only be 'an act of creative power'. Moreover, he was encouraged by Tait's experimental illustration of Helmholtz's theory by a demonstration of the properties of smoke rings. In response, Thomson formulated a theory of 'vortex atoms', an elaboration of his tentative speculation (which had seemed to have no empirical support) that a rotary motion within a continuous ether provided the basis for a theory of matter, and hence for a theory of the field. Thomson declared that 'Helmholtz's rings are the only true atoms', and his theory of vortex atoms posited a plenum filling space, action being transmitted along vortex filaments. The theory of a universal plenum, which considered atoms as discontinuities in the ether, thus solved the problem of the relationship between ether and matter and the problem of the mode of action in the physical field, the problems posed by Faraday with which Thomson had been grappling for more than twenty years. To provide a physical theory of the field Thomson required a mechanical theory of the ether, and the dynamical model he had postulated in his paper on magneto-optic rotation was justified by a physical theory of vortex motion. Action in the field was therefore propagated through an ethereal plenum, and the concept of a field of force could be represented in terms of the motion of an ethereal continuum.

Maxwell's Theories of the Field

The work of James Clerk Maxwell (1831–79) constitutes the most systematic attempt to develop Faraday's theories of the physical field. In his *Treatise on electricity and magnetism* (1873), Maxwell declared that he had translated Faraday's ideas into a mathematical form, but the relationship between Faraday and Maxwell cannot be construed simply as conceptual innovation followed by mathematisation. In providing a mathematical expression for Faraday's physical ideas, Maxwell transformed Faraday's concepts, formulating physical and mathematical models of the field that provided a more comprehensive physical theory and more profound conceptual insights into the problems raised by Faraday's ideas. The alternative modes of representation of the field proposed by Faraday – using the action of contiguous particles of an ambient medium and the theory of physical lines of force – were each subjected to development and comprehensive mathematical and conceptual analysis by Maxwell. While Thomson developed a theory of an ethereal continuum as a physical embodiment of Faraday's plenum of force, Maxwell explored the geometrical implications of the lines of force, and he also developed a physical model of ether particles to represent the transmission of action in the field by contiguous ether particles. Maxwell drew upon Thomson's mathematical and physical ideas, notably Thomson's use of the rotation of molecular vortices to represent the physical constitution of the field, employing models and methods introduced by Thomson in his attempt to achieve a representation of Faraday's physical concepts. Maxwell proposed a series of models, developing successive theories to avoid difficulties or inadequacies in the models previously formulated, each of which provided insights into the fundamental conceptual issue raised by Faraday: the relationship between the mode of propagation of force in the field and the nature of the material substance constituting the physical structure of the field.

Maxwell's philosophical sophistication was already apparent in his first theory of the field, published in 'On Faraday's lines of force' (1856). His introduction to Faraday's and Thomson's ideas and his first gropings towards a theory of the field can be traced from his letters to Thomson in 1854–5, and it is apparent that Maxwell was attracted to Faraday's theory of lines of force because of his own interest in geometry. Maxwell perceived that Faraday's concept of lines of force could be employed as a purely geometrical representation of the structure of the field. Maxwell was also interested in Thomson's early paper on the geometrical analogy between thermal and electrical phenomena. Thomson's 'allegorical representation' of electrostatics, as Maxwell described it, provided the paradigm for a theory of lines of force. Maxwell proposed to develop a geometrical representation of lines of force, a framework of propositions about lines and surfaces that could be applied to the representation of thermal or electrical phenomena but that were intended as a 'collection of purely geometrical truths'. Maxwell wished to emphasise the gap between mathematical and physical representation, but also to indicate the physical implications of a mathematical representation of nature.

In his paper 'On Faraday's lines of force', Maxwell elaborated a 'geometrical model' of the field, the direction of the forces acting in the field being represented by lines of force filling space. To explain the intensity of the force he supposed an incompressible fluid moving in tubes formed by lines of force. He indicated that this was not a physical representation of the field; the fluid was 'not even a hypothetical fluid', but 'merely a collection of imaginary properties'. He termed the geometrical model of fluid flow a 'physical analogy' that presented 'the mathematical ideas to the mind in an embodied form'. Although the concept of fluid flow was suggested by the mathematical analogy between the flow of heat and the

I have got a good deal of your ... electrical subjects, both Dewey(?) & through the famous G. Gibbs. I have also seen other helps, and now Faraday's three volumes.

My object in doing so very somewhat — someone to talk about down in electrical science, mathematically & experimentally, and to try to comprehend the same in a rational manner by the aid of any notions I could screw into my head. In nearly every case these notions have come upon you as I nearly may, which I have appropriated. Of these are Faraday's theory of polarity, & ... that perfect & every ... ... that perfect & every ... the whole others of action of the magnetic or electric bodies, also his "conducting power" of different media with the conducting power ... for them.

Then comes your ... physical representation of the case by ... by means of the conductors of heat, and your theorem that

$$\frac{d}{dx}\!\left(\alpha\frac{dV}{dx}\right)+\frac{d}{dy}\!\left(\alpha\frac{dV}{dy}\right)+\frac{d}{dz}\!\left(\alpha\frac{dV}{dz}\right)=4\pi\rho$$

Then Ampère's theory gives us galvanic about circuits, then form of your ... colours ...

incomparable elastic solid, Clerk ... the notion of the last demonstration in your R. S. paper on Magnetism. I have also seen numbers of Weber's theory of Electro Magnetism which are so ... of Electro Magnetism which I do not believe but speculation which I ought to be compared with other which ... to empound ... true ... & certainly gave many true ... & certainly gave ... at the expense of ... involving and ... ivory.

Now I have been planning & executing a system of ... which may be about lines of force in Electricity, then ... applied to Electricity, then afterward applied to Gplannum(?), & general Magnetism in Gplannum(?), & general Magnetism ... ... is in ... a collection of ... to think — real truths on ... conceptions of lines, surfaces ... sometimes ... The first part of my design is to form exceptions of lines, surfaces ... by ... that is not particularly ... the most important ... reasoning ... and ... about V ... about V ... equation in the last pages ($\alpha \times \beta ... + \gamma ...$) and to trace the lines of force ...

Fig. 4.3. Extract from a letter from Maxwell to William Thomson (1855). Like his fellow Scot Thomson, with whom he initiated a correspondence on the theory of electricity in 1854, Maxwell was a Cambridge-educated mathematician. The increasing application of mathematical analysis to physical problems was facilitated by the dominance of Stokes, Thomson, and Maxwell, who gave British physics its distinctive mathematical character. Although there were few chairs established in British universities in theoretical physics specifically, between 1850 and 1900 nearly half of the physics chairs in Britain were held by men educated in the Cambridge mathematical school.

As he remarked in this letter, Maxwell was interested in Faraday's physical ideas and in Thomson's 'allegorical representation' of electrostatics. Thomson's analogy between thermal and electrical phenomena. Maxwell perceived that the mathematical analogy between the equations of electrostatics and the equations for the flow of heat could be applied to a hydrodynamic analogy for electricity, a physical analogy for Faraday's concept of lines of force. Maxwell's sophisticated approach to the problem of analogy in physics shows evident traces of the influence of Scottish 'commonsense' philosophy, and of his education in philosophy at the University of Edinburgh. The Scottish philosophers, including Maxwell's Edinburgh professor, William Hamilton, stressed an abstractionist principle of knowledge involving comparison, a view echoed in Maxwell's discussion of physical and mathematical analogy. *Source:* James Clerk Maxwell to William Thomson (later Lord Kelvin), September 13, 1855 (Add 7342/M91, Cambridge University Library). Reproduced by kind permission of the Syndics of Cambridge University Library.

flow of electrical force, this geometrical representation did not have the status of a physical hypothesis; its use was grounded on 'mathematical resemblance', and its advantage over a purely analytical formalism was that the geometrical model of lines, surfaces, and tubes provided a visual representation of lines of force, keeping the phenomenon to be explained clearly in view. The mathematical analogy among fluid flow, heat, and electricity implied mathematical resemblance, not physical similarity; the similarity was between the mathematical relations, not the phenomena so related.

In adopting Faraday's theory of the primacy of lines of force, Maxwell emphasised the utility of Faraday's idea of lines of force as a geometrical representation of the spatial distribution of force in the field, rather than as a physical representation of the field. In a draft of the paper he observed, 'Faraday treats the distribution of forces in space as the primary phenomenon, and does not insist on any theory as to the nature of the centres of force round which these forces are generally but not always grouped'. Although he used lines of force to represent the field, he did not accept Faraday's physical interpretation of lines of force as depicting the field as a plenum of force, a theory of the field grounded on Faraday's concept of matter as filling space by its forces. In the *Treatise* Maxwell observed that Faraday had referred to the lines of force as belonging to the material substance of bodies, but he denied that this notion was essential to Faraday's theory of the field. For Maxwell, the theory of lines of force enunciated by Faraday represented the geometrical structure of the field, showing the spatial distribution, the direction, and the intensity of the forces, rather than implying that the physical structure of the field was to be envisaged as a plenum of force.

The geometrical focus of Maxwell's theory of lines of force can be seen from a letter he wrote to Faraday in 1857, in which he discussed the possibility of extending the concept of lines of force to explain gravitation. It was the geometrical imagery of Faraday's ideas, rather than the physical basis of his concepts, that attracted Maxwell's interest. 'The lines of Force from the Sun spread out from him and when they come near a planet *curve out from it* so that every planet diverts a number depending on its mass from their course and substitutes a system of its own so as to become something like a comet, *if lines of force were visible*'. The geometrical model of the field as represented by lines of force presupposed the assumption of absolute space. For Maxwell the concept of absolute space was the ground for charting the relations between material entities. The structure of space was not determined by the disposition of the field; the field, as represented geometrically by lines of force, was in space, and space was a fundamental ontological category, the condition for the existence of the field. In Maxwell's view a physical representation of the field required, as he put it in 'Faraday's lines of force', a 'mechanical conception' of the field, and in this paper he envisaged the mechanical representation of Faraday's concept of the electro-tonic state by means of a study of the laws of elastic solids, referring to Thomson's 1847 paper on the strain in an elastic solid as a model for the representation of electric and magnetic action. By the time Maxwell wrote to Faraday in 1857, he had become aware of the possibilities for a physical representation of the field implied by Thomson's explanation of magneto-optic rotation by the rotation of molecular vortices, and he drew Faraday's attention to Thomson's work, remarking that this theory appeared to provide a fruitful basis for the 'confirmation of the physical nature of magnetic lines of force'. For Maxwell a physical representation of the field required the formulation of a theory of the ambient ether as the substratum of the field.

Maxwell's intentions, foreshadowed in these remarks, were made explicit in the title of his paper 'On physical lines of force' (1861–2). In this paper he advanced from a discussion of the physical geometry of lines of force to a treatment of the electromagnetic field 'from a mechanical point of view'. He provided a systematic theory of the propagation of electrical and magnetic forces in terms of a stress in a 'magneto-electric medium'. He supposed that the magnetic field could be represented as a fluid filled with rotating vortex tubes, their geometrical arrangement corresponding to the lines of force, and the angular velocities of the vortices corresponding to the intensity of the field. He suggested a mechanical analogy to explain the rotation of vortices about parallel axes in the same direction: that of a machine in which an 'idle wheel' was placed between two wheels that were intended to revolve in the same direction. By analogy, he suggested that a layer of particles, acting as idle wheels, were placed between contiguous vortices so that the rotation of each vortex would cause the neighbouring vortices to revolve in the same direction. Maxwell represented the 'magneto-electric medium' as a cellular ether, depicted like a honeycomb. Each cell consisted of a molecular vortex surrounded by a layer of 'idle-wheel' particles that would be subjected to translational motion if adjacent vortices had different angular velocities, the motion of these particles corresponding to the flow of an electric current in an inhomogeneous magnetic field. The mechanical model thus had an electromagnetic correlative, and

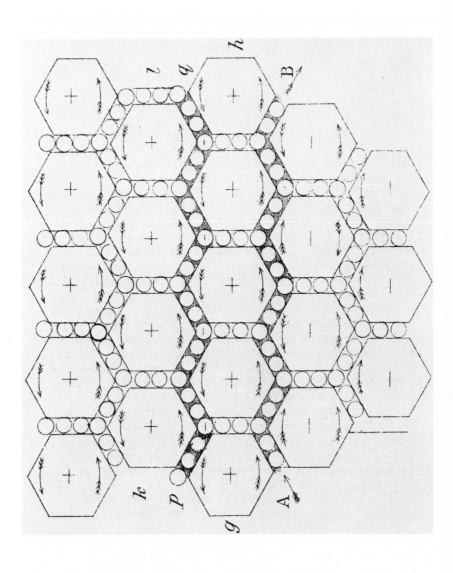

Fig. 4.4. Maxwell's physical model of molecular vortices and electric particles (1861). In this model of the ether Maxwell postulated a framework of vortices embedded in an incompressible fluid. Each vortex was separated from its neighbours by a layer of spherical particles, revolving in the opposite directions to the vortices. These 'idle-wheel' particles were identified with electricity. The electric current flowed from A to B, and the row of vortices gh above AB was set in motion in an anti-clockwise direction (+), and the row of vortices gh above AB was set in motion in an anti-clockwise direction (+), engaging the layer of particles pq, which acted on the next row of vortices kl. The transmission of electric action was explained in terms of the process of communicating the rotatory velocity of the vortices from one part of the field to another. The idle-wheel particles (electricity) permitted adjacent vortices to rotate in the same direction. The vortices below AB would rotate in a clockwise direction (−). The figure contains a drafting error: The vortices below AB would rotate in a clockwise direction (−), despite the directions on some of the arrows.

Maxwell emphasised that this model of idle-wheel particles was provisional. Nevertheless, he continued to argue that magneto-optic rotation implied that the rotation of vortices represented physical reality. This 1861 representation of the mechanical structure of the field was intended as a heuristic illustration, not a working mechanical device of the kind later invented by Boltzmann (and also by Maxwell himself), which provided specific mechanical analogues for electromagnetic phenomena (see Figure 6.1). *Source: The scientific papers of James Clerk Maxwell*, ed. W. D. Niven, 2 vols. (Cambridge, 1890), 1:488, pl. VIII, fig. 2.

provided a representation of electromagnetic induction. The vortex model was employed as a physical representation of the lines of force; abandoning the geometrical imagery of his earlier paper, Maxwell employed Faraday's concept of the electro-tonic state as a physical representation of the lines of force and provided a mechanical interpretation of the electro-tonic state as the rotational momentum of the vortices.

Maxwell consistently emphasised that his model of idle-wheel particles and a cellular ether was only a 'provisional and temporary' hypothesis. Confessing that the model might appear 'awkward', he made it clear that the supposition of a 'mechanically conceivable' model merely demonstrated the possibility of a mechanical explanation of the field, though the ether model he had suggested did not purport to be a 'mode of connexion existing in nature'. In a letter to Tait in 1867 he observed that his ether model was 'built up to show that the phenomena can be explained by mechanism. The nature of the mechanism is to the true mechanism what an orrery is to the Solar System'. In support of his contention that a mechanical representation was required, Maxwell consistently appealed to the magneto-optic effect, which, as he observed in the *Treatise,* indicated that 'some phenomenon of rotation is going on in the magnetic field' and which therefore required explanation 'by means of some kind of mechanism', though the hypothesis of molecular vortices was merely suggestive and illustrative.

Maxwell's reflections on the tentative and illustrative status of his ether model were intended as a conclusion to his paper; the extension of his theory to electrostatics, to a theory of electric charge, remained unstated. Moreover, the mechanism of the transmission of rotation from the idle-wheel particles to the cellular vortices was unclear. Further reflection on these problems led Maxwell to an elaboration of the theory of 'Physical lines of force' and resulted in the unexpected unification of optics and electromagnetism. Maxwell suggested that the cells of the ether were elastic and interacted elastically with the idle-wheel particles. The 'magneto-electric medium' was divided into cells separated by partitions formed of a layer of particles whose motions corresponded to an electric current; the motion of these electric particles distorted the cells and gave rise to an equal and opposite force originating from the elasticity of the cells.

As in Maxwell's earlier model, the mechanical structure of the ether had an electrical analogue. The elastic distortion of the cell had as its analogue a 'displacement' of electricity in the cell that represented the 'polarisation' of the particles of the magneto-electric medium. The incorporation of electric 'displacement' into Maxwell's equations of electromagnetism provided a symmetry between the equations of electricity and those of magnetism. Adopted as a representation of dielectric polarisation, the concept was sanctioned by the cellular ether model, but its status in Maxwell's formal equations of electromagnetism was ambiguous and its justification remained controversial. Nevertheless, Maxwell had resolved the problem of formulating a physical theory of the field by developing a mechanical representation of Faraday's particulate ether model – that is, his concepts of the electro-tonic state and the action of the contiguous parts of an ethereal medium – in place of the geometrical imagery of lines of force. The theory was a 'physical' theory of 'lines of force' because the field of force was represented mechanically by a theory of the transmission of force between the particles of the ether.

The supposition of the elasticity of the magneto-electric medium permitted the extension of the theory to electrostatics, with the elastic stresses of the mechanical ether corresponding to the electrostatic field, and thus completed Maxwell's comprehensive mechanical field theory. The theory had an unexpected implication. Maxwell calculated the velocity of transverse elastic waves corresponding to the propagation of an electric displacement in the magneto-electric medium and found that the transverse elastic waves were transmitted with the same velocity as light waves. He concluded, 'We can scarcely avoid the inference that *light consists in the transverse undulations of the same medium which is the cause of electric and magnetic phenomena'*. The identification of the electromagnetic and luminiferous media was a major development in the theory of the ether. Maxwell had formulated a mechanical theory of the ether that had optical and electromagnetic correlates. The unification of optics and electromagnetism by the mechanical theory of the ether was completed by a treatment of magneto-optic rotation in terms of the theory of molecular vortices. Maxwell's confidence in the physical existence of the ether was strengthened by the unification of optics and electromagnetism and by the explanation of magneto-optic rotation, but he continued to stress the hypothetical status of his ether model.

Maxwell's preoccupation with the status of his mechanical model of the field found expression in his 'Dynamical theory of the electromagnetic field' (1865), which was grounded on the methods of analytical dynamics employed by Green and MacCullagh in their dynamical theories of the ether, rather than on the construction of a specific mechanical model to represent the structure of the ether.

The change in perspective can be seen from his presentation of his 'electromagnetic theory of light', as he here explicitly termed it. Though the wave equation was now derived from electromagnetic laws, the ether in which the wave propagation occurred was still considered to be subject to the laws of dynamics, even though Maxwell abandoned the attempt to specify its mechanical structure.

There were two fundamental difficulties with the programme of formulating a mechanical model of the ether. Although Maxwell had emphasised the heuristic value of the idle-wheel model in providing a mechanical explanation of molecular vortices, there was the danger of confusing a hypothesis with reality, of supposing that the hypothesis itself was validated by the confirmation of conclusions drawn from it. Furthermore, there was in principle no limit to the number of mechanical models that could be proposed as hypotheses. Maxwell expressed the issue with characteristic clarity in the *Treatise:* 'The problem of determining the mechanism required to establish a given species of connexion between the motions of the parts of a system always admits of an infinite number of solutions'. He therefore proposed to abandon the mechanical analogy of a cellular ether so that the physical theory of the field could be formulated independently of the supposition of any specific mechanical model. The formulation of a physical representation of the field would nevertheless be constrained by the programme of mechanical explanation, for his dynamical theory of the electromagnetic field assumed that electromagnetic phenomena were produced by the motions of particles of matter, that action was transmitted in the field by 'a complicated mechanism capable of a vast variety of motion'. While abandoning any attempt to elaborate a specific model to describe this mechanism, he continued to employ mechanical terms, mechanical correlates of electromagnetic quantities, whose use was 'illustrative', not 'explanatory', highlighting the dynamical framework of the theory of the field.

The dynamical framework appears clearly in Maxwell's emphasis on the field as a repository of energy. Arguing that energy could exist only in connection with material substances, he concluded that the ethereal medium which constituted the electromagnetic field was the repository of the energy of the field. The complicated mechanism of the ether was subject to the laws of dynamics, and the field was represented dynamically as energy transformations in the ether. Whereas the use of mechanical correlates for electromagnetic quantities was merely illustrative, Maxwell noted that 'in speaking of the Energy of the field, however, I wish to be understood literally'. Because 'all energy is the same as mechanical energy', the energy in electromagnetic phenomena was to be referred to the kinetic energy of motion of the parts of the ether and the potential energy stored up in the connections of the mechanical structure of the ether, and the mechanism of the ether was subject to the general laws of dynamics. The propagation of electromagnetic waves was conceived as energy transformations in the ether.

In his *Treatise on electricity and magnetism* (1873), Maxwell developed this dynamical interpretation of electromagnetism, using the analytical formulation of dynamics articulated by Thomson and Tait in their *Treatise on natural philosophy* (1867). He represented the Lagrangian method as a formulation of generalised equations of motion considered as 'pure algebraical quantities' in a manner 'free from the intrusion of dynamical ideas' – a purely mathematical formalism that avoided reference to the concepts of momentum, velocity, and energy after they had been replaced by symbols in the generalised equations of motion. Maxwell, however, aimed to follow Thomson and Tait in seeking 'to cultivate our dynamical ideas'. By basing the mathematical argument on the concept of impulsive forces, this method 'kept out of view the mechanism by which the parts of the system are connected', while keeping 'constantly in mind the ideas appropriate to the fundamental science of dynamics', the dynamical concepts appropriate to the representation of physical reality. The method avoided explicit consideration of the mechanical connections of the electromagnetic field, though the dynamical basis of the mathematical formalism was constantly kept in view. This satisfied Maxwell's criterion for the acceptability of mathematical methods in physics: that any mathematical formalism must keep the physical problem clearly in focus. As in his use of the 'physical analogy' of fluid flow to represent the spatial distribution of lines of force, where the geometrical model of fluid flow provided a physical embodiment of lines of force, Maxwell's dynamical theory of the field underscored the link between the mathematical formalism and the physical reality depicted, ensuring that 'we must have our minds imbued with these dynamical truths as well as with mathematical methods'.

Maxwell emphasised the prominence that Thomson and Tait gave to the principle of the conservation of energy in abstract dynamics. He argued that the field was represented by the configuration and motion of a dynamical system, and followed Thomson and Tait in maintaining that the energy of the field could be specified without reference to the hidden structure of this dynamical system. The field was a structure of matter in motion, action

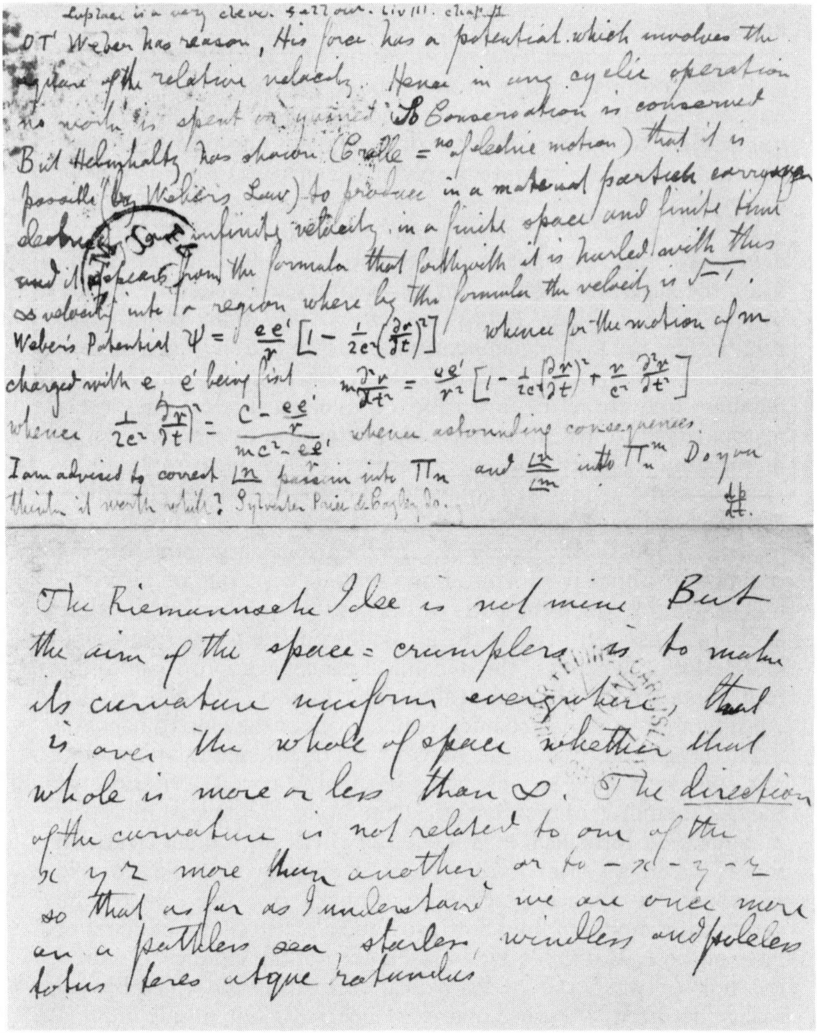

O T' Weber has reason, His force has a potential which involves the square of the relative velocity. Hence in any cyclic operation no work is spent or gained. So Conservation is conserved. But Helmholtz has shown (Crelle = no of electric motion) that it is possible (by Weber's Law) to produce in a material particle carrying electricity infinite velocity in a finite space and finite time and it appears from the formula that forthwith it is hurled with this ∞ velocity into a region where by the formula the velocity is $\sqrt{-1}$.

Weber's Potential $\Psi = \frac{ee'}{r}\left[1 - \frac{1}{2c^2}\left(\frac{dr}{dt}\right)^2\right]$ whence for the motion of m charged with e, e' being fixt $m\frac{d^2r}{dt^2} = \frac{ee'}{r^2}\left[1 - \frac{1}{2c^2}\left(\frac{dr}{dt}\right)^2 + \frac{r}{c^2}\frac{d^2r}{dt^2}\right]$

whence $\frac{1}{2c^2}\left(\frac{dr}{dt}\right)^2 = \frac{C - \frac{ee'}{r}}{mc^2 - \frac{ee'}{r}}$, whence astounding consequences.

I am advised to correct $|n$ passim into Π_n and $\frac{|n}{|m}$ into Π_n^m. Do you think it worth while? Sylvester Price &c.

$\frac{dp}{dt}$.

The Riemannsche Idee is not mine. But the aim of the space-crumplers is to make its curvature uniform everywhere, that is over the whole of space whether that whole is more or less than ∞. The direction of the curvature is not related to one of the x y z more than another, or to −x −y −z so that as far as I understand we are once more on a pathless sea, starless, windless and poleless totus teres atque rotundus

Fig. 4.5. Postcards from Maxwell to Tait (1871 and 1874). Tait was Maxwell's Edinburgh contemporary and closest scientific friend. In their correspondence, Maxwell and the two 'northern wizzards', as he described his Scottish colleagues William Thomson and Tait, joint authors of the *Treatise on natural philosophy* (1867), abbreviated their names. Thomson was *T*; Tait *T'*; and Maxwell became *dp/dt*, a reference to an equation in Tait's *Sketch of thermodynamics* (1868), *dp/dt* = *JCM*. This equation

being communicated between contiguous parts of the system. The magneto-optic effect indicated that the field was constituted of a material structure subject to rotation of its constituent parts, but the nature of the mechanism could be ignored. He stated that he had assumed the field to be a moving system, 'the motion being communicated from one part of the system to another by forces, the nature and laws of which we do not yet even attempt to define, because we can eliminate these forces from the equations of motion by the method given by Lagrange for any connected system'. Though changes in the structure and motion of the system led to the communication of action, the forces between different parts of the system were not defined, and the mechanical structure of the field was not specified.

Nevertheless, despite the power of this method, and even though Maxwell stressed the 'dynamical truths' that the formalism concealed and thus avoided the aridity of an entirely algebraic formalism, he remained dissatisfied with this mode of mechanical explanation. He pointed out that, according to the theory of the

---

Fig. 4.5 (*cont.*)

represented the second law of thermodynamics and was probably a deliberate notational joke by Tait: By using these symbols (Maxwell's initials) he gave his friend a thermodynamic signature.

In the first of these postcards Maxwell informed Tait that Helmholtz's 1847 claim that Weber's force law violated the law of energy conservation was invalid. In his papers on field theory Maxwell had upheld Helmholtz's argument, which was also repeated by Tait in his *Sketch of thermodynamics*. Maxwell went on to accept Helmholtz's new criticism of Weber's force law, however, a position he adhered to in his *Treatise on electricity and magnetism* (1873).

The second postcard was written in response to a query from Tait about Bernhard Riemann's paper on non-Euclidean geometry, which had been translated and published by William Kingdon Clifford (1845–79) in *Nature* in 1873. In indicating his disagreement with Riemann's work, Maxwell picked out the problem of the definition of coordinates in the theory, rejecting the concept of the curvature of space of the 'space-crumplers'. The Latin phrase ('the whole smooth and round') is a quotation from Horace. Maxwell was committed to the concept of absolute space and regarded the field as being *in* space, rather than conceiving the curvature of the lines of force as defining the geometric structure of space. *Source*: James Clerk Maxwell to P. G. Tait, November 7, 1871, and November 11, 1874 (Add 7655/Ib/36 and 72, Cambridge University Library). Reproduced by kind permission of the Syndics of Cambridge University Library.

Treatise, electrical action was 'a phenomenon due to an unknown cause, subject only to the general laws of dynamics', but that a 'complete dynamical theory' of the electromagnetic field would represent the hidden structure of the material system constituting the field, so that 'the whole intermediate mechanism and details of the motion, are taken as the objects of study'. Although he realised that an infinite number of mechanical ether models could be constructed so as to represent the field, his desire to achieve a 'complete' theory of the field led him to envisage the possibility of a mechanical model fully consistent with physical reality.

Ether and Field in British Physics, 1880–1900

In his later writings in the 1870s, Maxwell began to come to terms with the implications for the structure of the ether of Thomson's theory of vortex atomism. Maxwell observed that the simplicity of the fundamental assumptions of the vortex atom theory was a notable virtue. The inertia, density, and mobility of an incompressible universal plenum were sufficient to define the properties of the vortex atom without further arbitrary assumptions. Maxwell consistently emphasised that the phenomenon of magneto-optic rotation indicated that the ether had a molecular structure, but in the late 1870s he implied that molecular vortices in the ether could be discontinuities in an ethereal continuum, structures in a continuous ether endowed with the property of elasticity.

These remarks suggested a unification of Maxwell's theory of the electromagnetic field and Thomson's concept of a universal ethereal plenum. The theory of vortex atoms continued to be of interest as a theory of matter, and the field, considered as a plenum of force, came to be represented by an ethereal continuum rather than by action between contiguous ether particles. In his *Treatise on the motion of vortex rings* (1883), Joseph John Thomson (1856–1940) noted that the theory of atoms as vortex filaments provided a coherent atomic theory, and in his writings on field theory he proposed that the theory of tubes or lines of force required no additional assumptions about the material structure of the field, and was closer to ultimate physical reality than a theory of ether particles. John Henry Poynting (1852–1914) avoided using an ether model by employing the distribution of lines of force in space and a flux of energy between lines of force to represent the field. The electromagnetic field was envisaged as a homogeneous plenum, and electromagnetic phenomena were explained by the dispositions of lines of force.

In a paper entitled 'On the electromagnetic theory of the reflection and refraction of light' (1880), George Francis Fitzgerald (1857–1901) developed a theory of the ethereal plenum to incorporate Maxwell's electromagnetic theory of light. Whereas Maxwell had linked the electromagnetic variables to the material structure of the ether, Fitzgerald endowed the ether with purely electromagnetic properties. He revived the rotationally elastic ether proposed by his fellow Irishman MacCullagh in 1839, formulating an electromagnetic analogue of MacCullagh's theory of the optical ether. The rotation of MacCullagh's elastic solid was represented by Maxwell's electric 'displacement', and the velocity of ether streams was represented by magnetic force. Employing this electromagnetic analogue, Fitzgerald demonstrated that Maxwell's theory could be applied to the problem of optical reflection and refraction. Whereas Maxwell had emphasised that, to explain magneto-optic rotation, the material motion of the ether must be supposed, Fitzgerald wished to 'emancipate our minds from the thraldom of a material ether', and hence did not postulate the motion of the ether, merely characterising its electromagnetic properties. In later papers Fitzgerald proposed a theory in which ether and matter were represented by vortex motions in a universal plenum. Matter was a structure of closed vortex rings, and the ether was envisaged as vortex filaments stretching through space. The twists and waves in the vortex filaments gave rise to the propagation of electromagnetic forces, fulfilling the programme of explaining the physical structure of the field by a vortex ether. In 1885 Fitzgerald proposed a mechanical ether model composed of wheels and rubber bands; he intended merely to demonstrate that it was possible to represent the connection between ether and matter by means of a mechanism, and the model provided a mechanical illustration of ether strain, not an explanation of physical reality. Fitzgerald favoured the representation of the connection of ether and matter by vortex motions in a plenum rather than by appeal to an elastic solid theory of the ether – the conception of ether as 'a jelly with matter spread through it, like grapes in a jelly'. The crucial issue, as Fitzgerald noted, was the explanation of the interaction of ether and matter to explain magneto-optic rotation: 'The properties of a jelly prevent our supposing continuous rotation of its elements'.

In his *Baltimore lectures on molecular dynamics and the wave theory of light* (1904), delivered and issued in transcript in 1884, William Thomson elaborated an alternative approach to a theory of an ethereal continuum. As in the 1850s, the main problem facing

Thomson was the explanation of the relation between matter and the ether, and he attempted to discuss optical phenomena in relation to the elastic solid theory of the ether. Maxwell's demonstration of the electromagnetic theory of light argued that any ethereal medium representing the field should propagate transverse waves, and hence would have the properties of an elastic solid rather than of vortices in a fluid plenum. Moreover, Thomson found difficulties in demonstrating the stability of an elastic ether composed of vortex rings, and he abandoned the theory of vortex atoms. He aimed to formulate a 'comprehensive dynamics' embracing the ether, electromagnetism, and the wave theory of light. He considered a purely electromagnetic theory of light unacceptable, urging in its place the construction of a theory based upon 'plain matter-of-fact dynamics and the true elastic solid' as providing the only coherent foundation for the wave theory of light. Thomson emphasised that his models of the elastic solid ether were merely illustrative; all that could be said was that the ether behaved like an elastic solid in propagating light waves.

The central issue was Thomson's treatment of the relationship between ether and matter. He argued that optical phenomena had resisted satisfactory explanation in terms of the mechanical properties of an elastic solid luminiferous ether. To explain these phenomena he proposed a 'molecular' dynamics of ether particles, suggesting 'crude mechanical' models to explain the relation between the elastic solid ether and the particulate structure of matter and seeking to explore the way in which material particles acted on the ether. He considered the ether a 'continuous and homogeneous' elastic solid filling space, and he sought to investigate the nature of the 'elastic connection' between ether and the particles of matter. In place of his vortex atom theory of the ether plenum he investigated the mutual actions between a continuous elastic solid ether and the molecules of matter. He illustrated the phenomenon of optical dispersion by a model of rigid concentric spherical shells linked by springs to represent the vibrations of particles of matter in the ether. The phenomenon of magneto-optic rotation was explained by supposing the rotation of a gyroscopic molecule in the ether.

Thomson developed these ideas on a rotationally elastic ether in papers written in the late 1880s, elaborating his gyroscopic model and discussing the relations among the elastic solid ether, electromagnetism, and the structure of matter. To explain the elasticity of the ether he imagined the ether as a cellular structure of minute gyrostats; the rotational elasticity of the ether was accounted for by the inertia of the spinning motion of the gyrostats. Whereas the wave theory of light and electromagnetism pointed to an elastic ether subject to distortion, the magneto-optic effect indicated that the ether was rotational, a property most satisfactorily represented by vortices in a fluid. Thomson's attempt to unify elastic solid and rotationally elastic ether models by the gyrostatic ether was intended to resolve this difficulty. But a comprehensive ether dynamics eluded him. Although the vortex atom theory provided a model for the field as a plenum of force and represented the interaction between ether and matter, it failed to incorporate the electromagnetic theory of light. The mechanical models of the elastic solid ether theory failed to explain the relation between the mechanical properties of the ether and the 'molecular' dynamics of matter.

In an attempt to resolve the problems of constructing an ether theory that would represent all optical and electromagnetic phenomena, Joseph Larmor (1857–1942) published his systematic and synthetic 'Dynamical theory of the electric and luminiferous medium' (1893–7). Reviewing the first part of the paper for the Royal Society, J. J. Thomson observed that it was 'a kind of physical theory of the universe', a reflection of the monumental breadth of Larmor's work. Larmor proposed that the ether could be represented as a homogeneous fluid medium, if the ether was endowed with rotational elasticity that was latent and only manifested by a displacement of the elements of the ethereal medium. To represent the rotational elasticity of the ether he employed a mechanical model, William Thomson's model of spinning gyrostats. Larmor argued that the ether could be represented as a homogeneous fluid medium, unless it was subjected to elastic distortion. In this way his ether model embraced and synthesised MacCullagh's rotationally elastic ether (which Fitzgerald had transformed into an electromagnetic ether theory), William Thomson's theory that matter could be conceived as vortex rings in a primordial fluid medium, and Thomson's representation of the elasticity of the ethereal medium by the rotation of gyrostats. Larmor's ether had the properties of a perfectly incompressible and elastic fluid, and it was thereby able to be the medium for the propagation of transverse undulations.

Larmor's physical model of a fluid and rotationally elastic ether was intended merely as a 'working representation' of the ether, a strictly illustrative and heuristic model. Following Maxwell, Larmor emphasised the value of the Lagrangian formalism of dynamics in permitting the details of the mechanism of the ether to be ignored. His theory was grounded on an analytical function specifying the distribution of energy in the field. Larmor maintained that the

Lagrangian formalism was a sufficient explanation of physical reality, for it 'really involves in itself the solution of the whole problem'. All that could be known of the ether could be formulated as differential equations defining the properties of a continuous medium. He pointed out that his argument was largely independent of the rotational ether model, but observed that such hypotheses had heuristic value, providing a physical insight into the mathematical formalism.

Larmor stressed the difference between elastic transmission in the fluid ether and the elasticity attributed to ordinary material substances. His ether was a 'pure *continuum*', with elasticity, inertia, and the continuity of motion as its 'sole ultimate and fundamental properties', whereas ordinary matter was composed of discrete atoms endowed with inertia, so that the elasticity of matter arose from the interaction and distribution of the component particles. The properties of the ether were represented not by the configuration and interaction of ether particles, but by the elasticity of a homogeneous fluid plenum. Adopting the basic premise of the vortex atom theory, Larmor maintained that the ether was a primordial plenum, envisaging matter as a vortex structure in the ether. In his restatement of his theory in *Aether and matter* (1900), he observed that 'matter may be and likely is a structure in the aether, but certainly aether is not a structure made of matter'. The ether was physically 'prior to matter, and therefore not expressible in terms of matter'. The fundamental unit of matter was a centre of rotational strain in the vortex ether; Larmor supposed that these strain centres were endowed with an electric charge, and termed them 'electrons'. The electrons were the 'sole ultimate and unchanging singularities in the uniform all-pervading medium'. According to the electron theory of matter, atoms were represented by an aggregation of electrons describing stable orbits round each other. The electrons were nuclei of rotational strain in the ether, having a permanent existence in the ether just as a vortex ring possessed stability in a perfect fluid.

Larmor's theory of the ethereal continuum resolved two fundamental problems in the theory of the physical field: the relationship between ether and matter in the propagation of action in the field, and the relationship between the mechanical properties of the ether and the electromagnetic field. Action in the field was propagated by the rotationally elastic continuum, and the interaction between matter and the ether was explained by a theory of matter as centres of rotational strain in the ether. This was a dynamical theory of the field in that the fundamental properties of the ether were its inertia and elasticity, and the ether's mode of action was represented by the analytical formalism of Lagrangian dynamics. The electromagnetic properties of the field were explained by the theory of electrons, which were both centres of rotational strain in an ethereal plenum endowed with the fundamental properties of inertia and elasticity, and ultimate electromagnetic entities. Larmor's theory of the ethereal plenum unified the electromagnetic and dynamical properties of the ether, and thus synthesised the diverse strands of British theories of ether and field.

Force and Ether Theories in German Physics

The field and ether theories developed by British physicists in the nineteenth century constituted a radical departure from the form of electrical theory envisaged by Coulomb and Laplace, who posited a mathematical theory of electrical forces analogous to the central force law of gravity. The use of a mediating ether or field to explain electromagnetic phenomena, and the sophisticated treatment of mathematical and mechanical ether models by British physicists, transformed the theoretical presuppositions of electrical science. In Germany there was a less radical, but nevertheless marked, departure from the central force theory of electricity. Wilhelm Eduard Weber (1804–91) aimed to unify the phenomena of electricity and magnetism in a fundamental force law; he developed a theory of a particulate electromagnetic ether as a means of unifying electromagnetic and optical phenomena.

Weber's *Elektrodynamische Maassbestimmungen* [*Electrodynamical measurements*] (1846) introduced a departure from traditional central force theories of electricity. Weber's law for the forces acting at a distance between the particles of electric fluids included terms that made the force between two electrical particles depend not only on the inverse square of the distance between them (as in Coulomb's law for electrostatic forces), but also on their relative velocity and acceleration. Weber justified this modification of the traditional theory of central forces, which supposed a motion-invariant force law, by maintaining that his force law was merely a mathematical expression of electrical forces. Weber's force law did not provide an explanation of electrical forces, but merely a quantitative measure of the forces. Although Weber distinguished between the mathematisation of nature and the representation of physical reality, he also made it clear that his modification of traditional central force concepts was justified by his belief that electrical forces were propagated by an intervening medium analogous to the luminiferous ether. He indicated that he had in mind Ampère's theory of an electrical ether permeating space, on the supposition that the electrical ether was composed of positive and negative electrical fluids surrounding the molecules of ordinary matter. The additional terms in Weber's force law were therefore explained by interactions between the two electrical fluids, the velocity- and acceleration-dependent terms modifying the inverse-square law of electrostatic forces for electromagnetic and electromotive forces. Weber remarked that he envisaged an explanation of the undulatory theory of light in terms of the oscillations of the electrical ether, and he explained the magneto-optic effect by the rotation of the electrical ether.

The physical theory of the ether thus supplemented Weber's mathematical force law. It was in response to Weber's work that Gauss expressed his belief in the need to enunciate a 'constructible representation' of the propagation of electrical forces. Though Weber and Gauss noted that the mathematical formalism was independent of its physical embodiment, Weber was strongly committed to a physics based on the unifying role of an electrical ether. He developed this theory further in the 1850s and 1860s, making it apparent that his concept of force differed significantly from traditional central force theories. In Weber's theory, force did not simply act from one particle to another but linked two particles together to form an 'atom pair'; force existed as an embodiment of the relationship between the two particles.

In a paper published in 1871 Weber attempted to link his force law to energy concepts, expressing the force between two atoms as the potential energy of the 'atom pair'. He was replying to the criticism of his force law raised by Helmholtz in 1847: that, in supposing time-dependent terms in the electrical force law, Weber's law violated the principle of energy conservation. Weber did not reply to Helmholtz's criticisms until 1869, and though there was a protracted exchange of criticisms and replies during the 1870s, the delay in Weber's response had led Maxwell to adopt Helmholtz's critique of Weber's force law as an argument against any theory of electromagnetism based on a law of central forces, however modified. In responding to Helmholtz's arguments, Weber suggested that the concept of energy conservation required modification in its application to the theory of electrodynamics. He developed an electrical view of nature, subsuming chemical affinity, the theory of gases, and the luminiferous ether under a unifying theory of fundamental electrical particles. He argued that electric atom pairs could be formed of like as well as unlike particles. Ordinary matter consisted of negative particles closely bound together, whereas the ether was constituted of positive pairs loosely bound. Chemical combination was explained by the distribution of negative particles, and the properties of matter were held to depend on the nature of the connections between these particles and the electrical ether. The oscillations of the electrical ether explained the propagation of light waves by the transfer of energy.

Weber's modifications of central force concepts and his physical theory of the electrical ether led to the proposal of theories of the propagation of force through space. In the 1860s several papers were published that explored the possibilities of a mathematical description of the transmission of electromagnetic forces. Bernhard Riemann (1826–66) represented the transmission of electricity not instantaneously, as in a central force theory, but with the velocity of light in the luminiferous ether. The Danish physicist Ludwig Lorenz (1829–91) was influenced by Oersted in emphasising the unity of physical phenomena. Lorenz demonstrated that electrical forces were propagated at the velocity of light and concluded that the vibrations of light were electrical currents. Formulating a mathematical description of the propagation of electrical forces, he explained the transmission of force by field equations, regarding the propagation of force in the field as a matter of action between immediately adjacent elements of the field. Maxwell recognised that Lorenz's theory had formal similarities to his own field theory of electromagnetism and his electromagnetic theory of light. Lorenz maintained that light was a rotational vibration of electricity and envisaged the field as a material plenum subject to vibrations. Carl Neumann (1832–1925) explained the transmission of electrical action by reference to the propagation of energy; as Maxwell observed, this theory made no attempt to formulate a 'consistent [constructible] representation' of the propagation of energy in terms of 'the conception of a medium in which the propagation takes place'. Although these theories offered major departures from the central force theory of electricity (in supporting the hypothesis that the propagation of electrical action was time dependent and in supposing the mediating role of the ambient field), they differed from the dynamical explanations espoused by British physicists, which postulated mathematical and mechanical models of the ether as the material embodiment of the field.

These developments occurred independently of the introduction of field concepts into British physics, but in an influential paper on the theory of electrodynamics published in 1870, Helmholtz sought to achieve a synthesis of the innovations of Maxwell's electromagnetic theory of light and the formal consistency of a central force theory of electrodynamics, and to avoid the conceptual ambiguities that he believed threatened the coherence of Weber's electrodynamics. Helmholtz continued to reject Weber's theory, remaining convinced that Weber's time-dependent force law could not be satisfactorily reconciled with the law of energy conservation. Nor were the alternatives to Weber's theory proposed by Riemann, Lorenz, and Neumann acceptable to Helmholtz; in supposing the propagation of forces, these theories admitted time-dependent force laws. Helmholtz was fully aware that Maxwell's work provided a unified theory of electromagnetic phenomena, and that Maxwell's electromagnetic theory of light was a major conceptual innovation in physical theory. He attempted to formulate an action-at-a-distance theory of electrodynamics, consistent with his commitment to energy physics, that would incorporate the electromagnetic theory of the propagation of light without introducing a time-dependent force law.

Helmholtz derived a general formula describing the interaction between electric currents acting instantaneously at a distance, from which the competing laws of electrodynamics (such as Maxwell's and Weber's) could be deduced by mathematical manipulation. Helmholtz's formula did not introduce velocity-dependent force laws or suppose the propagation of force in time. However, by introducing limiting conditions, he derived a wave equation representing the transmission of states of polarisation through an electrically and magnetically polarisable medium. He established that a wave equation for electromagnetic propagation did not depend uniquely on the assumption of Maxwell's field theory. He assumed that the electromagnetic medium was subject to the electric and magnetic polarisation of its parts under the influence of electric and magnetic forces. The electric interaction of bodies was determined partly by direct action between distance forces and partly by the polarisation of the medium. The distance forces gave rise to forces acting at a distance between the polarised parts of the medium, and the polarised medium transmitted the forces directly at a distance between contiguous polarised parts. In this theory electric and magnetic forces were propagated instantaneously between polarised particles, not with the velocity of light, as in Maxwell's theory.

Helmholtz argued that the electromagnetic theory of light did not depend on the unique form of Maxwell's theory, but could be derived by introducing limiting conditions to the generalised force law, which was obtained from the traditional assumption of electrical action at a distance. It was through Helmholtz's version of the electromagnetic theory of light that Continental physicists became familiar with Maxwell's work. Maxwell's *Treatise* was considered to be a difficult and ambiguous work, and Helmholtz was thought to have remedied the conceptual impenetrability of Maxwell's ideas, making his theory intelligible. In the 1870s Hertz and H. A. Lorentz came to Maxwell's electromagnetic theory of light through Helmholtz's supposition of a central force theory. In presenting the electromagnetic theory of light as a deduction from his generalised force law, Helmholtz incorporated the electromagnetic theory of light within the accepted framework of electrodynamic theory, though he completely transformed the physical basis of Maxwell's theory in the process.

Hertz and Electromagnetic Waves

Helmholtz's arguments made a major contribution to the work of Heinrich Rudolf Hertz (1857–94). Hertz's famous experiments on the propagation of electromagnetic waves were conceived in response to a problem that Helmholtz had proposed when Hertz was his student, an experimental test of the relationship between polarisation and electromagnetic effects. Hertz made two fundamental contributions to the development of field theory: the direct verification of the propagation of electromagnetic waves, and a radical critique of the conceptual structure of the field equations in Maxwell's *Treatise*, which led to Hertz's reformulation of Maxwell's equations of electromagnetism. In both aspects of his work, the scientific problems that Hertz attempted to solve had been posed by Helmholtz, in one case in the search for an experimental test of Maxwell's work, and in the other in outlining the conceptual difficulties of Maxwell's electromagnetic theory.

In presenting evidence to support his electromagnetic theory in the *Treatise*, Maxwell had appealed mainly to optical phenomena, emphasising his derivation of the formula for the propagation of light. He proposed the electromagnetic ether as a medium for the propagation of light, and apparently did not contemplate the possibility of a direct experimental detection of electromagnetic waves. The propagation of electromagnetic vibrations at frequencies less than those of light, though reported in the literature, does not appear to have aroused his interest. Though the possibility of producing electromagnetic waves was implicit in Maxwell's theory, and had been discussed by Fitzgerald and Oliver Lodge (1851–1940), Hertz's impetus to perform the experiment came from Helmholtz.

Hertz's experimental vindication of Maxwell's electromagnetic theory developed from a theoretical examination of it inspired by Helmholtz's theory of electricity. In a paper published in 1884, Hertz explored the relations between field and central force theories of electrodynamics, demonstrating that Maxwell's assumptions about the propagation of electric force were consistent with the known laws of electrodynamics. Nevertheless, he refused to espouse a physical hypothesis about an electromagnetic ether, as Maxwell had done, being concerned merely to pinpoint the formal coherence of Maxwell's conception of the propagation of electric and magnetic forces. One feature of this paper is of especial importance for his later experimental work: his view that evidence for the propagation of electric and magnetic forces should be sought in electromagnetic rather than optical phenomena.

In November 1886 Hertz began the series of experimental studies that was to lead him to the confirmation of the propagation of electric waves in air. His first experiments followed the strategy of his 1884 conclusion that the propagation of electric action did not imply an electromagnetic ether; but by February 1888, when his seminal paper on the propagation of electromagnetic waves was presented to the Berlin Academy, the concept of an electromagnetic ether had become fundamental to Hertz's argument. He later observed that the hypothesis of a substance pervading air and empty space that was subject to electric and magnetic polarisation was the kernel of Maxwell's theory of the field. He began his paper by declaring that if polarisation effects existed in air, then 'electromagnetic actions must be propagated with finite velocity'. The detection of electromagnetic waves would confirm that there was a substance pervading air that was subject to electromagnetic polarisation, and would thus confirm the theory of an electromagnetic ether, the essential feature of Maxwell's field theory.

In a series of fundamental experiments, Hertz set himself the task of detecting electromagnetic waves and measuring their velocity. He produced electric waves with a wire connected to an induction coil, and detected them with a small loop of wire with a gap in which sparks could be detected when currents were induced. This apparatus enabled him to measure the wavelength of the electric waves, and with the calculated frequency of the oscillator he determined that the velocity of the electric waves was equal to the velocity of light. Hertz also explored the analogy between light and electric waves, focusing the rays with mirrors, demonstrating their reflection, refracting them through a prism of hard pitch, and demonstrating polarisation effects using metal gratings. He concluded that these experiments established the 'identity of light, radiant heat, and electromagnetic wave-motion'. Stressing that the experiments were valid independently of the assumption of any particular theory, he nevertheless made it plain that these experiments favoured Maxwell's theory of the electromagnetic ether. Hertz's experiments had a striking impact on Continental physicists, leading to the acceptance of the concept of the electromagnetic field. The Dutch physicist Henrik Antoon Lorentz (1853–1928), who had favoured Helmholtz's theory, signalled his acceptance of the theory of the electromagnetic field by declaring that Hertz's experiments were 'the greatest triumph that Maxwell's theory has achieved'.

Nevertheless, Maxwell's theory of electromagnetism posed considerable problems. In his 1884 paper Hertz had formulated field equations that expressed the propagation of forces in terms of formally symmetric relations between electric and magnetic forces, though without assuming any hypothesis about an electromagnetic ether. His new commitment to Maxwell's ether and his dissatisfaction with the formal coherence of Maxwell's *Treatise* led him to seek an axiomatic formulation of Maxwell's theory of the electromagnetic field. Maxwell himself had provided an analytical statement of field theory in a paper published in 1868, formulating fundamental equations of the electromagnetic field as theorems that expressed reciprocal relations between electric and magnetic forces. This succinct statement of field theory made no reference to an ether model, but provided an analytical basis for the field theory of electromagnetism. The use of symmetric relations between electric and magnetic forces in formulating the equations of the electromagnetic field was also proposed by Oliver Heaviside (1850–1925) in the late 1880s. In a formulation of the field concept that made no appeal to an ether model, Heaviside expressed the physical state of the electromagnetic field in terms of interactions between electric and magnetic forces and the flux of energy in the field. Despite notational differences, Hertz and Heaviside had elaborated similar formulations of the equations of the electromagnetic field. Hertz restated his equations in a paper on the theory of electrodynamics published in 1890, explicitly presenting his argument as an axiomatic formulation of Maxwell's theory of the electromagnetic field. Hertz expressed his ideas with a clarity

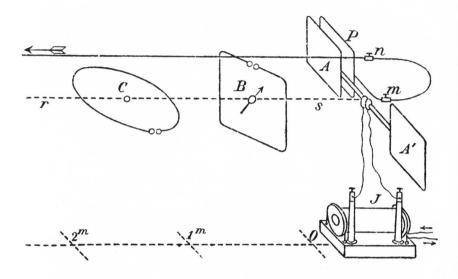

Fig. 4.6. Heinrich Hertz's experimental determination of the velocity of propagation of electromagnetic waves (1888). The apparatus produced electric waves in the spark gap between the copper plates AA' connected to an induction coil J, and the waves were detected by sparks gaps in square (B) or circular (C) loops of wire. To produce electric waves in a wire, Hertz placed a plate P behind the plate A and connected it to a wire mn carried out to an earth connection. Hertz demonstrated that there was interference between waves propagated directly through air and waves propagated through the wire, though his calculation of the velocity of electric waves in air was hampered by the difficulty of determining the velocity of propagation of waves in a wire. In a subsequent experiment on the reflection of electric waves, Hertz was able to measure the wavelength of electromagnetic waves in air independently of the propagation of waves in the wire. He thus demonstrated by experiment that electromagnetic actions were propagated through space with finite velocity, and established the agreement between the velocity of light and the velocity of electromagnetic waves. Hertz subsequently established the refraction and polarisation of electromagnetic waves, demonstrating the analogy between electromagnetic and optical waves. He maintained that his method provided experimental confirmation of Maxwell's electromagnetic theory of light. *Source:* Heinrich Hertz, *Untersuchungen über die Ausbreitung der elektrischen Kraft* (Leipzig, 1892), p. 116, fig. 25.

lacking in Heaviside's work, and he supplemented his mathematical argument with a justification of his approach to the theory of electromagnetism – and a critique of Maxwell – in the introduction to a collection of his papers on electricity, *Untersuchungen über die Ausbreitung der elektrischen Kraft* [*Investigations on the propagation of electric force*] (1892).

Hertz's argument focused on the ambiguities surrounding Maxwell's definition of electric charge and the concept of the 'displacement' of electricity, which Maxwell had introduced as a representation of the polarisation of the ether. Hertz was troubled by the relation between the polarisation of the ether and the displacement of electricity. His solution to the problem was to eliminate displacement as a concept independent of the electric force, hence avoiding the necessity of explaining whether a displacement of electric charge was the cause or the effect of polarisation. In Hertz's statement of the equations of the electromagnetic field, displacement was eliminated in favour of a formulation based on symmetrical relations between the electric and magnetic forces. Hertz claimed that these equations expressed the essential part of Maxwell's work, and he declared that 'Maxwell's theory is Maxwell's system of equations'. However, Maxwell's equations of the electromagnetic field in the *Treatise* embodied the concepts that Hertz had eliminated, so Hertz's assertion more accurately expresses the view that the essential part of Maxwell's theory was defined by Hertz's formulation of the equations of the electromagnetic field.

Hertz's axiomatic formulation of the equations of the field avoided the 'concrete representations' of electromagnetic concepts employed by Maxwell. Although Hertz wished to distinguish between the 'gay garment' of a mechanical model and the formalism of field theory, he did not renounce the mechanical view of nature. He remained committed to the belief that the electromagnetic waves were produced by an ether whose parts were connected by a mechanical structure. Nevertheless, the distinction he drew between the formalism of electromagnetism and its representation by a mechanical model was influential; it gave the theory of electromagnetism a clarity of expression lacking in Maxwell's *Treatise* and highlighted the central problem of formulating a consistent interpretation of the equations determining electric and magnetic forces. Hertz's simplification of the formalism of the electromagnetic field equations was a major contribution to the acceptance of field theory.

The Problem of Motion through the Ether

Hertz's brilliant experimental demonstration of electric waves was seen as establishing the fundamental tenet of Maxwell's theory, the concept of the electromagnetic ether. Addressing the British Association in 1888 on the subject of Hertz's work, Fitzgerald declared that 'Hertz's experiment proves the ethereal theory of electromagnetism'. Nevertheless, the physical status of the ether and the relationship between the theory of the luminiferous ether and the concepts of electromagnetism were the subject of considerable debate during the 1880s, a debate that Hertz joined in 1890 with a theoretical paper on the electrodynamics of moving bodies.

Hertz based his theory of the electrodynamics of moving bodies on the postulate that the ether was mechanically dragged by ponderable bodies moving through it. He argued that there was no contradiction between this assumption and known electromagnetic phenomena and that by making this assumption he was able to develop a systematic formulation of electrodynamics consistent with the theory he had established for the electrodynamics of bodies at rest. He indicated that this assumption was inadequate in explaining optical phenomena, however, observing that a fully comprehensive theory would distinguish between the conditions of the ether and the matter embedded in it. To attempt to formulate such a theory would require the admission of arbitrary hypotheses unwarranted by the problem under considerations, the representation of the electrodynamics of moving bodies. In stressing the limitations of his theory, Hertz underscored the problem of the relationship between optical and electrodynamic phenomena, and indicated that the preservation of Maxwell's unification of optics and electromagnetism in electromagnetic ether theory required the elaboration of a systematic representation of the motion of bodies through the ether, and hence an account of the relationship between ether and matter.

The optical phenomena that stood outside Hertz's electrodynamics included stellar aberration and ether drag. The phenomenon of aberration – the apparent displacement of a star as the result of the motion of the earth – presented difficulties for the theory of the luminiferous ether. Explanation of stellar aberration by the wave theory of light seemed to require the assumption that the ether was not disturbed by the passage of the earth through it, as any disturbance of the ether would deflect the light rays from their rectilinear path. To avoid the difficulties that would ensue if it were supposed that the ether was disturbed by the earth's motion, both

Young and Fresnel accepted the hypothesis that the ether freely pervaded the substance of all material bodies, and that the ether was not disturbed by the motion of the earth through it. The main issue in the discussion of aberration and ether drag in the early nineteenth century concerned the implication of stellar aberration for the nature of the luminiferous ether. To explain the failure of an optical experiment performed by Arago to detect the relative motion between the ether and the earth, Fresnel suggested that transparent bodies had partial ether drag effect on the light passing through them; his calculation of this effect was confirmed in 1851 by Hippolyte Fizeau (1819–96).

The Young–Fresnel theory of the unhindered passage of the earth through the ether was, however, questioned by Stokes in 1845. Stokes conceived the ether as a viscous elastic solid, concluding that there must be friction between the ether and the earth moving through it. The unhindered passage of the earth through the ether, as envisaged by Fresnel, was therefore firmly rejected. In place of Fresnel's theory of the stationary ether, Stokes maintained that the earth and planets dragged along with them the ether that was close to their surfaces; beyond the boundary of ether drag, the ether in space was undisturbed by the motion of the earth. To explain stellar aberration Stokes provided an analysis of the motion of the earth through the ether, imposing a mathematical restriction on the motion of the ether which had the physical consequence that there would be no eddies in the ether flowing round the earth. Although Fresnel's theory explained aberration and the question of the earth's motion through the ether equally well, Stokes had provided an alternative account of the problems, a theory that explained the earth's passage through the ether according to his influential elastic solid model of the ether.

The problem of ether drag attracted attention from a different perspective in the 1880s, following the assimilation of Maxwell's demonstration that the ether played a role in the phenomena of electromagnetism. The problems of the relation between ether and matter and the influence of the motion of the earth on electromagnetic phenomena were brought into focus by Maxwell's work. Maxwell had himself discussed the experimental detection of the earth's motion through the ether, suggesting that the ether could possibly be detected by measuring the variation in the velocity of light when light was propagated in opposite directions. Maxwell's remarks stimulated the American physicist Albert Abraham Michelson (1852–1931) to an experimental test of the variation in the velocity of light under the circumstances discussed by Maxwell.

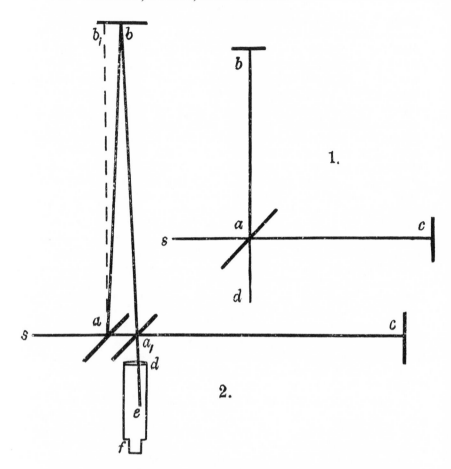

Fig. 4.7. The experiment to detect the relative motion of the earth and the luminiferous ether by A. A. Michelson and E. W. Morley (1887). In Figure 1, *sa* is a ray of light that is partly reflected along *ab*, partly transmitted along *ac* by a semitransparent mirror, and reflected by the mirrors *b* and *c* along *ba* and *ca*; *ba* is partly transmitted along *ad*; and *ca* is partly reflected along *ad*. If the paths *ab* and *ac* are equal, the two rays will interfere along *ad*. Figure 2 illustrates the arrangement if the apparatus moved in the direction *sc*, with the velocity of the earth in its orbit, the ether being assumed to be at rest. The direction and distances traversed by the rays would be altered, and the difference in the paths would be in the opposite direction if the whole apparatus were turned through ninety degrees. Michelson and Morley sought to measure the displacement of the interference fringes when the apparatus was moved through ninety degrees and to

Michelson was possibly aware of Fizeau's use of a semitransparent mirror to send a beam of light over a distance and back to an observer behind the mirror, and he conceived the idea of letting two light beams from the same source simultaneously run a double journey along two paths of the same length but placed at right angles to each other. Michelson designed an apparatus (an interferometer) to measure the interference pattern produced by the two light beams. Anticipating the confirmation of Fresnel's theory of a stationary ether, which had been supported by Fizeau's experiment, Michelson supposed that the two light beams would be differently affected by their motion through the ether, as a result of their different orientation to the direction of the earth's motion through the stationary ether. The experiment was performed in 1881 and yielded the surprising result that the earth's motion relative to the ether was undetectable. Michelson concluded that the hypothesis of a stationary ether was incorrect, indicating that Stokes's theory of an ether that was dragged by the passage of the earth through it was to be preferred.

In 1886 Lorentz published a paper on the influence of the motion of the earth on optical phenomena, focusing on the relation between ether and matter. Lorentz criticised Michelson's calculations, and also argued that Stokes's theory of a dragged ether was based on inadmissible assumptions. Lorentz's arguments favoured Fresnel's hypothesis of a stationary ether. Though this was the theory that Michelson had expected to confirm, the result of his

---

Fig. 4.7 (*cont.*)
calculate the fringe shift to be expected on the basis of Fresnel's theory of stellar aberration.

Michelson's interferometer, a semitransparent mirror (which divided the source ray of light *sa* into two beams) and a telescopic arrangement to focus and measure the interference fringes, was a delicate and carefully constructed piece of apparatus. The experiment was performed with the greatest care to eliminate vibrations, by mounting the apparatus on a massive stone floating on mercury. The apparatus was covered with a wooden box to eliminate air currents and changes of temperature. Contrary to their expectations, the observed shift of the interference bands when the apparatus was moved through ninety degrees was so small that Michelson and Morley concluded that if there was any relative motion between the earth and the luminiferous ether, it must be so small as entirely to refute Fresnel's explanation of aberration. *Source*: A. A. Michelson and E. W. Morley, 'On the relative motion of the earth and the ether', *American Journal of Science*, 3d ser. 34 (1887), 335, figs. 1 and 2.

experiment had placed Fresnel's theory in doubt. To resolve the issue Michelson, in collaboration with Edward Williams Morley (1838–1923), performed two further experiments, repetitions of Fizeau's test of Fresnel's theory of partial ether drag and of Michelson's ether drift experiment. Michelson and Morley established that Fizeau's confirmation of Fresnel's hypothesis of partial ether drag was correct, and from this they concluded that Fresnel's theory of the stationary ether was confirmed. The result of their repetition of Michelson's ether drift experiment implied a contrary conclusion, however: Once again the experiment yielded (within the limits of experimental error) a 'negative' result, failing to detect the earth's motion through the ether. Michelson and Morley concluded that if there was any relative motion between the earth and the ether, it must be so small as to refute the theory of the stationary ether.

When Hertz discussed the electrodynamics of moving bodies in 1890, the problems of ether drag and of the connection between ether and matter were under active consideration by physicists. Although Lorentz's theoretical analysis and the repetition of Fizeau's experiment favoured Fresnel's hypothesis of the unhindered passage of the earth through the ether, the Michelson–Morley ether drift experiment of 1887 had concluded decisively against Fresnel's stationary ether. In supposing that the ether was dragged by the passage of the earth through it, Hertz was consciously making a simplifying assumption, leaving to subsequent analysis the task of establishing a treatment of electrodynamics that was based on a physically convincing theory of the relationship between ether and matter. Maxwell, Lorentz, and Hertz all emphasised the fundamental importance of formulating a theory that would establish the relation between ether and matter. As Hertz observed, the establishment of such a theory would provide the basis for a theory of the electrodynamics of moving bodies that was grounded on coherent physical assumptions.

Lorentz and the Electromagnetic World View

Lorentz's systematic treatment of the relation between the ether and matter and the electrodynamics of moving bodies was formulated in the 1890s in the context of his electromagnetic theory of nature, which supposed that electric particles (successively termed 'ions' in 1895 and 'electrons' in 1899) were embedded in an electromagnetic ether. In his mature theory of the electromagnetic ether he proposed to divest the ether of all mechanical properties, reducing the laws of nature to properties defined by the electromagnetic field equations. Lorentz's theory developed a universal physics grounded on electromagnetic concepts. In elaborating this electromagnetic world view, Lorentz was able to explain the relationship between ether and matter. Unlike Maxwell, who considered the ether to be a state of matter, Lorentz completely separated the two. Conceiving matter as charged particles (electrons), he distinguished matter from the ether and explained the connection between ether and matter in terms of the relationship between electrons and the electromagnetic ether, hence separating the electromagnetic field from matter.

Lorentz's electromagnetic world view originated in his studies of the ether and the electromagnetic theory of light. In his doctoral thesis of 1875 he employed Helmholtz's derivation of the electromagnetic wave equation, which he applied to an analysis of optical reflection and refraction; and he continued to espouse Helmholtz's approach to electrodynamics and the electromagnetic theory of light until 1891. Convinced that Hertz's experimental demonstration of electromagnetic waves constituted a convincing confirmation of Maxwell's concept of the field, Lorentz declared his acceptance of Maxwell's field concept. Nevertheless, Lorentz disagreed with Hertz's theory of electrodynamics. In his first paper on the electron theory, 'La théorie électromagnétique de Maxwell et son application aux corps mouvants' ['Maxwell's electromagnetic theory and its application to moving bodies'] (1892), Lorentz criticised Hertz's purely axiomatic statement of the field equations and the assumption of a totally dragged ether in Hertz's account of the electrodynamics of moving bodies.

Emphasising his commitment to the theory of the stationary ether, Lorentz sought to establish electromagnetic theory by using the Lagrangian formulation of dynamics that Maxwell had employed in the *Treatise*. Lorentz argued that the ether was completely separate from ordinary matter, and the theory of electrons enabled him to establish the connection between ether and matter. He argued that molecules of ordinary matter contained positive and negative charged electrical particles (electrons). The electromagnetic field resulted from the motions of these particles, and the field could act on ordinary matter by exerting forces on the electrons embedded in matter. Lorentz's electrodynamics used electric particles to provide an explanation of electric charge and to formulate a consistent theory of the electromagnetic field. Lorentz viewed his theory as a synthesis of Continental electrodynamics, which explained electrical action by the forces exerted between particles, and Maxwell's concept of the electromagnetic field, which emphasised that electric action was propagated at the speed of light. Lorentz's electrodynamic equations characterised the electromagnetic field and the connection between the field and the electrons. His paper included a demonstration that Fresnel's theory of partial ether drag by a moving transparent body could be interpreted as indicating the interference of light rather than a mechanical dragging of the ether. Lorentz argued that the effect of a moving transparent body on light passing through it was due to the interference between the incident light and the light waves produced by the vibrations of the electrons, vibrations produced by the effect of the incident light on the electrons.

Lorentz thus affirmed the theory of the stationary ether. Nevertheless, the stationary ether hypothesis was in conflict with the 'negative' result of the Michelson–Morley experiment, and in a paper written later in 1892 Lorentz advanced the suggestion (which was also proposed independently by Fitzgerald) that the interferometer arms contracted in the direction of the earth's motion through the ether. He interpreted the contraction effect by supposing that the molecular forces that determined the dimensions of bodies were propagated through the ether like electric forces. The contraction effect and other compensatory actions eliminated the effects to be expected if the ether was not dragged by the motion of the earth through it, thus explaining the 'negative' result of the Michelson–Morley experiment.

Despite Lorentz's emphasis on the 'dynamical' foundation of his electrodynamics, his theory did not posit a mechanical connection between the ether and ordinary matter. The forces between the ether and the electrons were electrical, and his theory implied an electromagnetic rather than a mechanical view of nature. His major treatise, *Versuch einer Theorie der electrischen und optischen Erscheinungen in bewegten Körpern* [*Inquiry into a theory of electrical and optical phenomena in moving bodies*] (1895), renounced the appeal to mechanical principles in deriving the basic equations of the theory. Following a method similar to Hertz's, Lorentz stated the electromagnetic field equations and the equation connecting the field to the electrons as fundamental electromagnetic postulates. Unlike Hertz, who remained committed to the theory of the mechanical ether, Lorentz abandoned the framework of dynamical principles in favour of an explicitly electromagnetic theory of nature, in which electrodynamics rather than mechanics was envisaged as providing the conceptual foundations of physics. Lorentz established a more exact mathematical form of the contraction hypothesis, repeating his argument that the molecular forces that determined the dimensions of the interferometer arm were propagated through the ether like electric forces, an argument that was now presented in the context of his electromagnetic theory of nature. His dissent from the mechanical view of nature was further emphasised by his explicit demonstration that the field was denuded of all mechanical properties; separated from ordinary matter, the field was an independent physical reality.

In subsequent elaborations of his theory, Lorentz further modified the traditional mechanical foundation of physical theory. The empirical discovery of the discrete unit of electricity (the 'electron') seemed to Lorentz to provide empirical confirmation of his theory, and he developed a theory in which the properties of electrons were conceived in nonmechanical, electromagnetic terms. Electromagnetism was envisaged as providing the conceptual foundations of physics. Gravitation was explained by the theory of the electromagnetic ether, and the laws of mechanics were viewed as special cases of universal electromagnetic laws. Lorentz defined inertia and mass in electromagnetic terms, and he denied the constancy of mass, a fundamental principle of Newtonian mechanics. Lorentz's theory of mass as an electromagnetic concept was a striking and influential feature of this theory.

Around 1900, Lorentz's theory exerted a profound influence on the development of physical theory. Many physicists argued that electrodynamics, rather than mechanics, would provide the unifying conceptual foundations of physics. The concept of the ether, denuded of all mechanical properties, seemed to many physicists to provide the basis for all physical theory. An ontology of electrons and the electromagnetic ether, not based on a framework of mechanical principles, was being developed in place of the mechanical concepts that had dominated physical theorising in the nineteenth century. In reformulating the theory of the electromagnetic field, Lorentz introduced a major departure from the programme of mechanical explanation. Whereas Hertz continued to affirm that the electromagnetic ether should be represented in terms of the motion of hidden masses, Lorentz maintained that the postulates of field theory should be based on an electromagnetic ontology. Whereas Larmor's theory of the electromagnetic field conceived a dynamical theory of an ethereal plenum and postulated electrons as centres of rotational strain in the ether, Lorentz rejected the dynamical foundations of the theory of the field and envisaged the foundation of physics on purely electromagnetic concepts.

CHAPTER V

Matter Theory: Problems of Molecular Physics

The physical constitution of matter appeared uncertain in the nineteenth century. Although an ontology of particles of matter in motion was fundamental to the programme of mechanical explanation and to the conceptual coherence of the science of thermodynamics, physicists were careful to distinguish between the general supposition of a particulate theory of matter and the adoption of more specific models of molecular structure. Though the mechanical view of heat as the motion of the particles of matter underlay the principle of the equivalence of heat and work, physicists found compelling evidence for a molecular theory of matter only with the development of the kinetic theory of gases in the 1850s. But the problem of explaining the phenomena of spectroscopy indicated the need to suppose complex internal molecular vibrations, and raised difficult questions about the formal coherence of the kinetic theory of gases. The problems of molecular physics raised crucial issues about the conflicting empirical constraints (from spectroscopy and the kinetic theory of gases) on the formulation of a coherent theory of the molecular structure of bodies. The problems of molecular physics shaped the development of thermodynamics: The statistical theory of molecular motions, which was formulated as a seminal feature of the kinetic theory of gases, led to the interpretation of the second law of thermodynamics as an irreducibly statistical law. For chemists, the problems of matter theory seemed equally complex, and the status of the atomic theory remained the subject of debate.

Chemical Atomism

The chemical 'atomic' theory employed by nineteenth-century chemists displays significant differences from the atomistic chemistry formulated by Newton, which had played a significant role in shaping the development of chemical theory in the eighteenth century. In Newtonian chemistry, a chemical element possessed its properties because of the way in which atoms formed an arrangement of particles that was characteristic of that element, these particles then forming further aggregations. All chemical elements were therefore envisaged as differently compounded structures of ultimate atoms. The atoms were held together by interparticulate forces, and Newtonian chemists proposed a quantitative science of chemical affinities. Other chemists, however, questioned the applicability of this theory to the study of chemical substances and reactions, following G. E. Stahl in maintaining a distinction between physical and chemical principles, a distinction between the supposition of physical atomism (which Stahl maintained was unable to explain chemical phenomena) and a theory of chemical elements as the last products of chemical analysis. Stahl's emphasis on a universal chemical philosophy influenced French chemists and shaped Lavoisier's definition of chemical elements as the final results of chemical analysis. Lavoisier noted that although the study of chemical affinities was in principle amenable to the application of mathematics, the subject lacked secure foundations as the basis for chemical theory; and though he accepted the atomic theory as a representation of the physical constitution of bodies, he considered speculation about atoms to be irrelevant to chemical science.

The work of John Dalton (1766–1844) provided a theoretical basis for chemistry congruent with this definition of the chemical elements. Dalton conflated the traditional distinction between ultimate physical atoms and chemical elements in his theory of 'chemical atomism'. In his *New system of chemical philosophy* (1808–27), he identified the diversity of chemical elements with heterogeneous chemical 'atoms', rejecting Newton's distinction between primordial physical atoms and the heterogeneous chemical elements composed of aggregations of these atoms. Dalton also abandoned the Newtonian search for force mechanisms in favour of a system of quantification based on the relative *weights* of chemical atoms. The origins of Dalton's chemical atomic theory lay in his studies on the mixing of gases and the absorption of gases in water, in which he rejected the explanatory role of forces of chemical affinity. In his theory of the absorption of gases he emphasised the importance of the relative weights of the ultimate particles of substances in determining the relative solubility of different gases in water. The attempt to calculate relative atomic

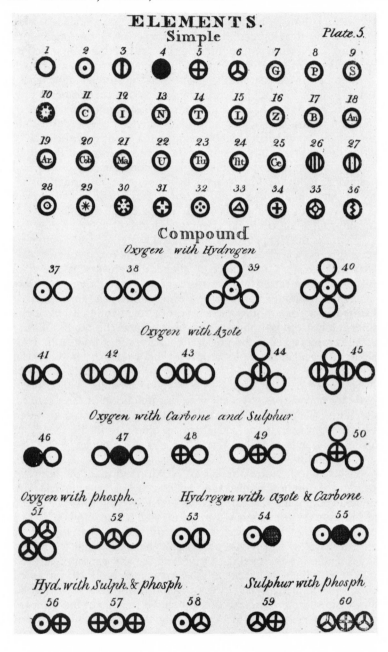

weights led his interest to shift gradually towards chemistry and the problems of chemical combination.

In the *New system* Dalton incorporated the law of equivalent chemical proportions formulated by Jeremias Benjamin Richter (1762–1807). In Richter's theory of 'stoichiometry', chemical combination was represented by the relative weights of elements calculated directly from analytical data. Following Richter, Dalton proposed that the chemical elements combined in simple integral ratios, and he asserted that chemical atoms combined in fixed proportions. The chemical atomic theory provided an explanation for the analytical ratios of stoichiometry, but Dalton's term 'atom' was used not only in the sense of the smallest particle of a chemical element possessing a given chemical nature and entering into chemical combination with other elementary particles in integral multiples, but also in the sense of solid and indivisible ultimate particles. Dalton thus drew no distinction between the chemical elements and primordial physical atoms.

Fig. 5.1. Table of the elements, from Dalton's *New system of chemical philosophy*, pt. 2 (1810). This plate displays Dalton's list of symbols for the elements, slightly revised from the symbols he used in the first part of the *New system* (1808). An accompanying table listed the weights of the atoms of the elements, relative to hydrogen. The first ten elements and their relative weights were oxygen (7), hydrogen (1), nitrogen (5), carbon (5.4), sulphur (13), phosphorus (9), gold (140?), platinum (100?), silver (100), mercury (167). Chemical compounds were represented as arrangements of atoms: Figure 37 represents water; Figure 54, ammonia; and Figure 56, hydrogen sulphide. These diagrams displayed Dalton's belief that the chemical elements were composed of heterogeneous atoms. Dalton's chemical atomic theory proposed that chemically indivisible particles (atoms) possessed unique atomic weights; but Dalton also asserted that the 'atoms' corresponding to the chemical elements were the fundamental particles in nature.

Dalton's calculations of relative atomic weights was grounded on his assumption that molecules of chemical substances consisted of the simplest possible combination of their component atoms. He justified this rule of simplicity by arguing that the fewer like, and therefore mutually repellent, atoms there were in a molecule, the greater the internal stability of the molecule. This argument was derived from his discussion of the mixing of gases, where he had stressed the mutual repulsion between gas particles of a similar kind. *Source*: John Dalton, *New system of chemical philosophy*, pt. 2 (Manchester, 1810), p. 548, pl. 5.

Dalton's quantification of chemistry was based not on the mathematical description of the forces of chemical affinity, but on the establishment of rules of chemical combination and on the classification of chemical atoms according to their relative weights. The formation of chemical compounds from the combination of elements in fixed proportions by weight thus implied that the relative weights of their atoms were in the same ratios. Although Dalton, in the first volume of the *New system,* stressed the utility of relative particle weights and the rules of chemical combination for chemical analysis, he also claimed a privileged status for his chemical atoms as ultimate particles in nature. In the second volume of the *New system* (1810) and in later statements on chemical atoms, he affirmed the physical reality of the different chemical atoms as fundamental particles, denying the Newtonian doctrine of the complex internal structure of the chemical elements. The diagrams in the *New system* displaying ponderable, indivisible atoms surrounded by atmospheres of caloric emphasised his belief in solid, spherical atoms as the ultimate chemical and physical particles in nature.

The distinction between 'chemical' and 'physical' atomism is of fundamental importance in charting the impact of Dalton's ideas on nineteenth-century chemistry. Physicists and chemists recognised a distinction between the status of the chemical atomic theory, which provided the conceptual rationale for the system of relative 'atomic' weights and for the assignment of molecular formulae to chemical substances, and the assumption of the physical reality and nature of atoms. The diagrams and models employed by chemists to represent molecular structures were generally interpreted as symbolic representations of the geometrical structure of molecules. To mitigate the physical implications of Dalton's theory, chemists sought to avoid assumptions about atoms in formulating the laws of stoichiometry, while maintaining the system of relative particle weights. Between 1810 and 1814, Davy, William Hyde Wollaston (1766–1828), and Jöns Jacob Berzelius (1779–1848) formulated tables of relative particle weights that purported to avoid atomistic assumptions. Davy argued that stoichiometry, rather than speculations about the ultimate particles of matter, constituted the kernel of Dalton's theory. He proposed a system of 'proportions' or relative weights as the basis for the calculation of the definite proportions in chemical combination. Wollaston's similar table of 'chemical equivalents' provided a system of elemental weights – a calculus of stoichiometrical quantities that was proposed as a practical tool in the calculation of chemical proportions and was widely used on the grounds of its empirical basis. Nevertheless, Wollaston ultimately admitted that the intelligibility of his system required the assumption that equivalents expressed relative atomic weights. Berzelius's systematic treatment of the laws of stoichiometry was initially published as an empirical table of combining ratios, though on reading Dalton's *New system* he produced a table of elemental weights and molecular formulae.

There was considerable agreement among these chemists in their treatment of stoichiometry and elemental weights. The stoichiometrical systems of Berzelius, Davy, and Wollaston were similar to Dalton's, apart from Dalton's use of the conceptually loaded term 'atom'. Reviewing the issue in his *History of the inductive sciences* (1837), William Whewell observed that the term 'atom' was more convenient than the alternative terms 'proportion' or 'equivalent' in denoting relative atomic weights, but he emphasised that the use of 'atom' need not be understood as implying assent to the hypothesis of ultimate, indivisible particles. By the 1830s the use of the 'atomic' theory in its narrowly chemical sense, to denote chemically indivisible particles, was well established and provided the basis for the discussion of molecular formulae that had become the primary concern of chemists. The status of the atoms themselves and the symbolism of molecular formulae continued to be debated. In 1837 Jean Baptiste Dumas (1800–84) remarked that he wished to erase the term 'atom' from chemical science on the grounds that it went beyond experience. He contested the assumption of the indivisibility of 'atoms' in Dalton's theory, but not the supposition of chemical particles that provided the rationale for discussing the molecular groups constituting chemical compounds.

The relationship between organic and inorganic compounds became a major concern of chemists in the nineteenth century. Berzelius argued that the exploration of the properties of organic compounds should be based on the established principles of inorganic chemistry. He maintained that a theory of chemical atomism must explain the combination of atoms in specific proportions in terms of a theory of chemical affinity, and he supposed that the electrochemical polarity of the atoms of compounds could explain the laws of stoichiometry. Chemical combination involved the neutralisation of the positive and negative electrical charges of individual atoms or of aggregates of atoms, a theory of combination that Berzelius sought to apply to the study of organic compounds. He stressed the importance of the arrangement of atoms in chemical compounds. In response to the synthesis of urea from ammonia and cyanic acid by Friedrich Wöhler (1800–82) in 1828,

Berzelius argued that Wohler's transformation of an inorganic salt (ammonium cyanate) into an organic compound (urea) provided an example of a phenomenon that he termed 'isomerism', in which substances of a similar composition, having the same chemical constituents in the same proportions, had dissimilar molecular configuration, which Berzelius explained by his electrical theory of chemical affinity.

By the 1830s chemists were especially concerned with deducing the arrangement of elements within organic compounds once their elemental composition had been established – that is, with discovering the structural disposition of atoms within compounds. The 'radical' theory of Liebig and Dumas postulated analogies between inorganic elements and 'radicals' such as the methyl and ethyl radicals, composite groups of elements within organic compounds, bringing organic reactions within the framework of the laws of inorganic combination. By the middle of the nineteenth century, chemical atomism was deeply embedded in the theory of organic chemistry. August Laurent (1807–53) employed ideas drawn from crystallography in representing the arrangement of atoms in organic compounds by three-dimensional geometrical structures. Though Laurent questioned the assumption that the molecular constitution of organic compounds could be inferred from their chemical reactions, he nevertheless emphasised the construction of models of molecular structure. Laurent stressed that these structures were symbolic formulae that expressed the analogies between compounds of similar constitution, but did not represent the real distribution of atoms in organic compounds. Organic chemists employed models to represent the molecular structure of organic compounds. As August Kekulé (1829–96) observed, writing in 1867, the programme of chemical atomism was based not on the supposition of the actual existence of atoms considered as indivisible particles, but on the application of atomism to the explanation of chemical reactions and the structure of chemical compounds.

Despite the establishment of chemical atomism as a fundamental precept of chemical theory, the status of chemical atoms was questioned by Benjamin Collins Brodie (1817–80) in 1866. Brodie, who proposed a chemical symbolism in which the laws of chemical combination were classified in a manner independent of any atomistic interpretation, was influenced by the work of Charles Frédéric Gerhardt (1816–56) in organic chemistry. Gerhardt had maintained that chemical formulae should merely represent the propensity of compounds to undergo reactions. Questioning the assumption that molecular constitution could be inferred from chemical reactions, he abandoned the attempt to represent molecular constitution; formulae described the disposition of a compound to display chemical properties. Brodie's chemical calculus sought to represent chemical formulae by the sequence of operations required to produce chemical substances, and he proposed a chemical algebra in place of the geometry of molecular structures. Brodie's work provoked a lively debate. Alexander William Williamson (1824–1904) responded by arguing that the issue concerning the status of the chemical atomic theory was not the indivisibility of atoms, an issue that was not raised in chemistry, but the representation of chemical phenomena by means of a corpuscular language that would represent atomic weights and molecular formulae. In Williamson's view the phenomena of chemistry demanded interpretation according to the chemical atomic theory; and Kekulé insisted that the assumption of chemical atomism was essential for the explanation of chemical phenomena.

Nevertheless, by the 1860s, the chemical atomic theory was firmly established. In 1860 a conference of chemists was convened at Karlsruhe to discuss the status of the theory, and Stanislao Cannizzaro (1826–1910) distributed a paper in which he revived a hypothesis first proposed by Amedeo Avogadro (1776–1856) in 1811. According to 'Avogadro's hypothesis', equal volumes of gases, under the same conditions of temperature and pressure, contain equal numbers of molecules. Though this hypothesis was widely familiar, Avogadro's work had also provided a method of defining molecular formulae and relative atomic weights. The weights of the molecules of gases were proportional to the densities of the gases, and the molecular formulae of compounds could be deduced from the volumes of the reactants. This method of defining atomic weights had little impact on chemical debate until Cannizzaro's demonstration that Avogadro's work provided a method of obtaining a system of atomic weights and reinforced the conceptual status of the chemical atomic theory. In 1860 Maxwell had given, as a derivation from his theory of gases, a mathematical demonstration of the chemical law that equal volumes of gases contained an equal number of particles. As Maxwell emphasised to the chemists during the discussion of Brodie's work, the kinetic theory of gases provided important evidence of the particulate theory of matter. Although the nature of atoms remained a matter for speculation, by the 1860s the molecular theory of matter and the chemical atomic theory were incorporated into the fabric of physical science.

Molecular Physics: The Kinetic Theory of Gases

When in 1850 Clausius established the conceptual foundations of thermodynamics, he noted that although the principle of the equivalence of heat and work could be seen to be conceptually intelligible if it were assumed that heat consisted in the motion of the constituent particles of bodies, he avoided consideration of the nature of the molecular motions that gave rise to the phenomena of heat and that were convertible to mechanical work. Instead, he formulated the axioms of thermodynamics independently of hypotheses about the nature of matter. Clausius's paper 'Ueber die Art der Bewegung welche wir Wärme nennen' ['On the kind of motion which we call heat'] (1857) provided a full account of a theory of molecular motions, defining the framework of gas theory in relation to the fundamental problem of thermodynamics, the relationship between thermodynamic concepts and the structure of matter.

Although not the first 'kinetic' theory of gases, Clausius's work was the first systematic treatment of the subject. The theory that the properties of gases were to be explained by the motion of gas particles had been advocated in the eighteenth century, notably by Daniel Bernoulli, but the 'kinetic' theory of gas particles in motion was never a serious rival to the Newtonian 'static' theory of gases, in which gas particles were considered to be stationary and held in place by interparticulate repulsive forces. By the 1770s the static theory was being elaborated in relation to the prevailing theory of imponderable fluids, the repulsive forces between gas particles being attributed to the presence of the imponderable fluid of heat, which was later termed 'caloric'. The caloric theory of gases remained the dominant view of gas structure until the 1850s, when the caloric theory of heat had been displaced by wave or mechanical theories of heat, although kinetic theories were proposed by John Herapath (1790–1868) and John James Waterston (1811–83). In demonstrating the equivalence of heat and work, Joule espoused Herapath's view of the motion of gas particles, and he argued that heat was to be attributed to the motion of the particles of bodies. Clausius drew similar conclusions from the establishment of the equivalence of heat and work, and stimulated by a paper by August Krönig (1822–79), which discussed the problem of treating the motions of gas particles by probabilistic arguments, Clausius published his first paper on the kinetic theory of gases in 1857.

Clausius had already employed probabilistic arguments in a discussion of meteorological optics. He had explained the blue colour of the sky as the reflection of light from particles floating in the atmosphere, and had used a statistical argument to justify his use of mean values in describing the motions of the particles. His commitment to the theory that heat consisted of molecular motions, and his use of probabilistic arguments in describing the motions of particles, provided a framework of interests that were stimulated by Krönig's paper. Clausius assumed that gas particles could be regarded as elastic spheres and observed that the duration of particulate collisions would be negligible and that the effect of intermolecular forces could be ignored in treating the motion and collision of gas molecules. Arguing that the temperature of a gas could be expressed in terms of the energy of the gas molecules, he emphasised that his value for the velocity of the molecules was an average velocity. The velocities of individual molecules could differ from this mean value.

This recognition of the relevance of a probabilistic argument in the theoretical treatment of the motion of gas molecules was taken further in a subsequent paper. In response to criticism that his theory was unable to explain the slow diffusion of gas molecules, Clausius introduced a more sophisticated molecular model to explain the motion and interaction of gas molecules. He stressed the role of intermolecular forces in determining the collisions of gas particles. Drawing on traditional ideas about molecular forces, he argued that molecular forces were attractive at large distances and repulsive when the molecules were in close proximity; the distance at which the attractive and repulsive forces of a molecule balanced was the radius of the 'sphere of action' of the molecule. The sphere of action determined whether, in molecular encounters, interaction between molecules occurred through reciprocal attraction or impact. Clausius introduced an important new concept, the 'mean free path', the distance that, on the average, a molecule would move before its centre of gravity came within the sphere of action of another molecule, at which point molecular collision would result. He was able to obtain an expression for the ratio of the mean free path to the sphere of action, but not to obtain values for these magnitudes. Treating the motion of the gas molecules as a random process, Clausius used a probabilistic argument in discussing the encounters between molecules.

Maxwell brought a greater mathematical sophistication to the analysis of the variations in the velocity of the molecules in a gas. Whereas Clausius had simply taken the average molecular velocity, Maxwell saw that a statistical analysis of the distribution of velocities was required. In Maxwell's treatment of the problem in his

'Illustrations of the dynamical theory of gases' (1860), the velocities were distributed among the molecules in accordance with a distribution function, a method similar to Laplace's analysis of the distribution of errors. Maxwell's theory was a dynamical theory of gases or, as he later described it in revising his terminology under the influence of Thomson and Tait's *Treatise on natural philosophy* (1867), a 'kinetic' theory of gases, in being concerned with the motions and collisions of particles. Maxwell's interest in the molecular theory of gases was stimulated by Clausius's papers, though Clausius's work fell on prepared ground. Maxwell's interest in the mathematical analysis of particle collisions and in the use of probabilistic arguments had already been aroused. In an essay on the rings of Saturn, in which he argued that the rings were a collection of small particles, he had commented on the difficulty of describing the complex mathematical laws of collisions among a multitude of particles; as he later observed, Clausius's work provided a general method for the analysis of the collisions of particles. Maxwell was also aware of the interest among physicists and mathematicians in the 1850s in the logical and mathematical status of probability theory, and his prior interest in probabilities and the mathematical analysis of particle collisions explains the appeal of Clausius's papers. With his characteristic emphasis on the method of physical analogy, Maxwell declared that he intended to arrange his treatment of the collision of particles in a manner independent of speculations about gases. The paper originated as a mathematical exercise, from which the mathematical propositions on the collisions of particles could be applied to the study of the properties of gases, just as his geometrical imagery of lines of force could be applied to the phenomena of electromagnetism. Although Maxwell's work began as a mathematical treatment of particle mechanics, in the published paper he presented a physical theory of the molecular properties of matter.

The focuses of Maxwell's paper were on the discussion of molecular collisions (on the assumption that gas molecules could be considered as elastic spheres) and the derivation of equations for the transport processes of gases, viscosity, heat conduction, and diffusion, from his calculation of the mean free path of gases. His subsequent experimental demonstration that the viscosity of gases varied directly with temperature contradicted his elastic sphere molecular model, leading him to reconsider his approach to the problem of molecular structure. His refinement of his theory in his 'Dynamical theory of gases' (1867) was engendered by Clausius's criticisms of his assumptions in deriving the velocity distribution function and of his derivation of the properties of gases from the concept of the mean free path. Maxwell abandoned the concept of the mean free path as a means of analysing molecular motions, and considered encounters between molecules envisaged as 'centres of force', rather than collisions between elastic spheres. He found that if the force law of repulsion between the centres of force was an inverse fifth-power law, the viscosity of gases varied directly with temperature, in agreement with his own experiments. In his new derivation of the velocity distribution law he discussed the effect of molecular collisions on the distribution law, determining the equilibrium distribution of molecular velocities, the distribution that would be unchanged by molecular collisions.

Maxwell's work on gas theory had a dual emphasis: on the properties of molecules and on the mathematical analysis of molecular motions. In a lecture entitled 'Molecules' (1873), he surveyed the range of evidence that provided information about the nature of matter, observing that the kinetic theory of gases provided both evidence that the properties of matter could be explained by a theory of molecular motions and a framework for the formulation of models of molecular structure. Moreover, spectroscopy indicated that vibrating molecules were the source of spectra, and spectroscopy could also provide the basis for a theory of molecular structure. The conflicting constraints of gas theory and spectroscopy were to shape subsequent work on molecular physics.

Maxwell distinguished the theory of gases, which was based on the 'statistical' method of explanation, from a theory based on the 'dynamical' method, concerned with the motion of individual particles of matter. He argued that the theory of gases could not be subjected to the dynamical or 'historical' method because its application required perfect data about the motions of individual molecules. As he explained in the lecture 'Molecules', the smallest portion of matter amenable to experiment consisted of millions of molecules; the motions of individual molecules were unobservable. Experiments in molecular physics could not 'pretend to absolute precision'; they provided only 'statistical information' about large aggregates of molecules. Maxwell maintained that in the theory of gases the probability, or moral certainty, of the regularity of averages replaced the 'absolute' certainty of the strict dynamical method: Despite the 'stability of the averages of large numbers of variable events . . . in a particular case a very different event might occur' from that expected from the 'regularity of averages', though 'we are morally certain that such an event will not take place'. He

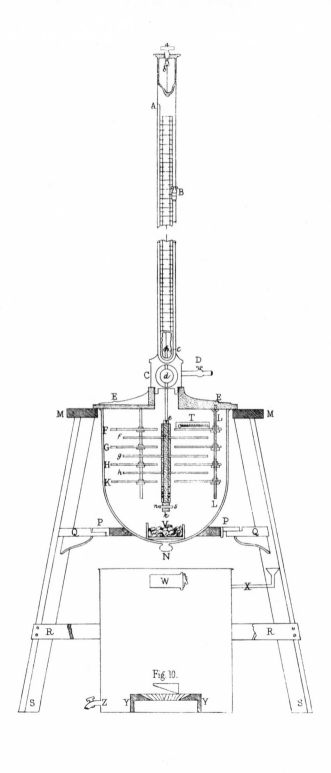

Fig. 10.

therefore sought to distinguish the theory of gases, as a theory based on the statistical method, from a theory of individual molecular motions. It was from his reflections on the status of the theory of molecular motions as an irreducibly statistical theory that Maxwell developed his argument that the second law of thermodynamics was an essentially statistical law, and could not be expressed as a dynamical law that would describe individual molecular motions.

Problems of Molecular Structure

Maxwell distinguished problems in molecular science from the discussion of the nature of 'ultimate atoms'. In his article 'Atom' (1875), written for the ninth edition of the *Encyclopaedia Britannica*, he observed that molecular science was concerned with the properties of matter and with the evidence for its particulate structure, whereas a satisfactory atomic theory of matter would be

---

Fig. 5.2. The apparatus used by Maxwell and his wife to determine the viscosity of gases as a function of pressure (1866). The experiment consisted in observing the oscillations of glass discs f, g, and h suspended in a sealed chamber. The pressure of the gas in the vessel could be varied and read on the barometer ACB. The temperature of the vessel could be varied by filling the tin vessel (Figure 10) with hot water, steam, or cold water, and raising it so as to envelope the chamber; and the temperature could be read on the thermometer T. By attaching a small piece of magnetised steel wire ns to the axis to which the glass discs were attached, and placing magnets under N, Maxwell set the discs in motion.

In his 1860 paper Maxwell had derived equations for the transport processes of gases, viscosity, heat conduction, and diffusion, and his experimental confirmation of his prediction that the viscosity was independent of pressure confirmed the hypothesis of a kinetic theory of gases. The relationship between viscosity and temperature was, however, dependent on the molecular model assumed. Maxwell had established that for a gas of elastic-sphere molecules, the viscosity could vary as the square root of the temperature. His experiments demonstrated that the viscosity varied directly with temperature, a result that contradicted the hypothesis of elastic sphere molecules. These experiments led Maxwell to reconsider the nature of the molecular structure of gases, and he abandoned the elastic sphere model of his 1860 paper in favour of a theory of molecules as 'centres of force'. *Source*: Maxwell, *Scientific papers*, 2:24, pl. IX, figs. 1 and 10.

required to provide an explanation of mass and gravitation. He discussed the vortex atom theory as an atomic theory, as a possible complete theory of substance that would provide an explanation of the immutability of matter and the cause of gravity. By contrast, the kinetic theory of gases provided evidence of the particulate structure of matter and of the properties and motions of molecules. Maxwell argued that the kinetic theory of gases provided the basis for the formulation of models of the molecular structure of substances.

The problem facing the molecular theory of matter, as Maxwell realised, was the conflicting constraints imposed by different molecular phenomena on the formulation of models of molecular structure. Maxwell had derived from his kinetic theory of gases the mathematical result that the kinetic energy of gas molecules was distributed equally among the internal mechanical motions of the molecules. This 'equipartition theorem' of the equalisation of energy was in conflict with experimental determinations of thermal properties (the specific heats) of gases, which implied a restriction in the internal motions of molecules. The limitations on the mechanical properties of molecules required to explain the structure of gas molecules were in conflict with the evidence of spectroscopy, which indicated that spectral lines were the result of complex internal molecular vibrations and contradicted the supposition of restrictions on the internal motions of molecules. There were, therefore, inconsistencies in the molecular theory of gases itself, and conflicts between the evidence drawn from gas theory and spectroscopy, in formulating molecular models adequate to the demands of theory and experiment.

Maxwell was committed to the equipartition theorem; he explored ways of explaining the properties of gases by discussing molecular models in which the internal mechanical motions were restricted, but recognised the difficulty of resolving the problem. Despite the acknowledged problems associated with the equipartition theorem, many physicists used it in discussing the internal structure of molecules. A variety of molecular models were proposed in an effort to vary mechanical motions in a manner consistent with the properties of gases, and the theory recorded some successes. Experiments on mercury vapour demonstrated that mercury was monatomic, in agreement with Maxwell's theoretical arguments; and in the 1890s John William Strutt, Lord Rayleigh (1842–1919), established that argon was a monatomic gas, using the equipartition theorem in support of this conclusion. Nevertheless, physicists recognised the difficulties in formulating satisfactory models of molecular structure that agreed with experimental results.

Spectroscopy posed a major challenge to the equipartition theorem. By the 1860s discussions of the nature of spectra were seen to bear upon the problems of molecular physics. The occurrence of line spectra, dark lines in the solar spectrum, had been discussed from early in the century and had been described in detail by Joseph Fraunhofer (1787–1826). The relationship between the 'Fraunhofer' line spectra and the flame spectra produced when chemical substances were introduced into flames had also been the subject of much discussion. In the late 1850s Robert Bunsen (1811–99) and Gustav Robert Kirchhoff (1824–87) had charged the characteristic spectral lines of different chemical elements. In a paper published in 1859 Kirchhoff established the relationship between the dark and bright lines in line and flame spectra. Arguing that chemical elements could both emit and absorb light of the same wavelength, he suggested that the dark lines in the solar spectrum were the result of the selective absorption of light from the sun by elements in the solar atmosphere. Spectroscopy became of interest to chemists, providing as it did a method of chemical analysis, and provoked speculations on the nature of the chemical elements.

Throughout the nineteenth century, hypotheses on the complexity of the elements were widely canvassed by chemists. William Prout (1785–1850) had suggested that hydrogen was the 'first matter' from which all other elements were composed, and had claimed that the atomic weights of all the elements were integral multiples of the atomic weight of hydrogen. Although it became apparent that this correspondence was only an approximation, Prout's ideas remained the subject of debate. In the 1870s and 1880s, speculations were made in the context of the discussion of the chemical implications of spectroscopy. Analysis of stellar spectra revealed the prevalence of the lighter gases in the spectra of the hottest stars, whereas cooler stars contained a greater proportion of heavier metals; this analysis led to hypotheses on the evolution of the chemical elements from simpler substances as a result of the cooling and condensation of the lighter elements in stars. Though discussion of the evolution and complexity of the elements remained speculative, spectral analysis was considered to be an important technique in the study of chemical elements, as well as a provider of insights into molecular structure.

Maxwell perceived the implications of spectroscopy for molecular physics and argued that spectra were the result of molecular

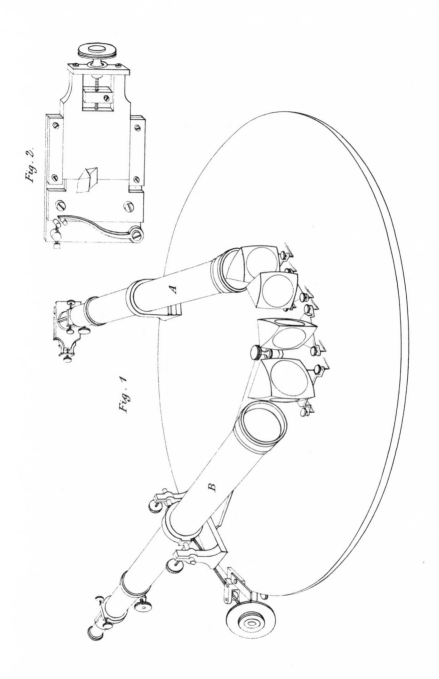

Fig. 2.

Fig. 1

A

B

Fig. 5.3. Kirchhoff's apparatus for the examination of spectra (1861). With this spectroscopic apparatus, Kirchhoff was able to compare the dark 'Fraunhofer' lines in the solar spectrum and the spectra of the elements. The sun's spectrum was observed through the telescope and prism arrangement illustrated in Figure 1, the eye piece of telescope A being replaced by a narrow slit and a divided scale being attached to the micrometer screw by which telescope B was moved so as to measure the distances between the spectral lines. So that the flame spectra of the elements could be compared directly with the Fraunhofer lines, Kirchhoff attached a prism arrangement (top right) that permitted light from flame spectra to enter the upper half of the slit while sunlight entered the lower half. Kirchhoff investigated the coincidence of bright lines in the sodium spectrum and two very strong lines (denoted D in the alphabetic scheme) in the solar spectrum. He argued that sodium selectively emitted lines of the wavelength of the D lines, and also tended to absorb selectively incident light of the same wavelength. He suggested that the dark D lines in the solar spectrum were produced by the selective absorption of light from the hot nucleus of the sun by sodium in the solar atmosphere. The dark lines in solar and stellar spectra provided evidence of the chemical composition of the solar and stellar atmospheres. *Source:* Gustav Robert Kirchhoff, *Untersuchungen über das Sonnenspectrum und die Spectren der chemischen Elemente* (Berlin, 1861), pl. III.

vibrations being communicated to the ether. Supposing that molecules vibrated like bells, he suggested that the aural harmonics of the bell were analogous to the spectral lines produced by vibrating molecules. In the 1880s physicists developed molecular models to explain spectra, arguing that spectra would provide a guide to the internal vibrations of molecules. It was realised that molecular models that were sufficiently complex to produce the vibrations necessary for spectra were incompatible with the equipartition theorem, which implied a restriction in the mechanical structure of molecules. Conversely, molecular structures that were compatible with the equipartition theorem were too simple to produce the lines observed in the spectra of the chemical elements. In response to this difficulty, Larmor, Fitzgerald, and Arthur Schuster (1851–1934) proposed electromagnetic theories in which spectra were explained by means of the oscillations of electrons.

The equipartition theorem itself was subjected to close analysis. William Thomson sought to demonstrate that it was invalid, in an attempt to preserve the molecular theory of matter. The formulation of mechanical models of molecular structure, under the guidance of the conceptual framework of the kinetic theory of gases, seemed to Thomson to be threatened if the equipartition theorem was valid. In response, Rayleigh maintained that the theorem was a necessary consequence of the kinetic theory of gases. Ludwig Boltzmann (1844–1906) argued that the equipartition theorem was a fundamental feature of the kinetic theory of gases, and he continued to elaborate molecular models consistent with the theorem in an attempt to avoid the difficulties posed by the properties of gases. The equipartition theorem remained an essential feature of his view of molecules as mechanical systems.

Discussion of molecular forces provided a way of avoiding the problems posed by the equipartition theorem. In 1870 Clausius, attempting to explain the second law of thermodynamics in terms of mechanical principles, had derived a mechanical theorem that related heat and temperature. This 'virial theorem' related the kinetic energy of molecules to the forces acting on them, and in 1873 Johannes Diderik van der Waals (1857–1923) applied the virial theorem to the study of the properties of gases. Maxwell saw that van der Waals's study of molecular forces could be developed into an investigation of the properties of matter, one that would stimulate further discussion on molecular forces and molecular structure and that would also avoid reference to the equipartition theorem.

The problems of molecular physics remained unresolved, for

though physicists adopted alternative approaches in an attempt to avoid the contradictions imposed by the conflicting constraints of spectroscopy and the kinetic theory of gases, the status of the equipartition theorem remained uncertain. In 1900 William Thomson described it as 'a cloud which has obscured the brilliance of the molecular theory of heat and light'. While Thomson concluded that the simplest way to preserve the coherence of the mechanical view of nature would be to reject the equipartition theorem, Boltzmann argued that the theorem was an essential part of the kinetic theory of gases, which established the conceptual coherence of the mechanical view of nature.

Molecular Physics and Thermodynamics

Clausius had formulated his kinetic theory of gases to provide a mechanical basis for thermodynamics, and it was from consideration of the implications of the statistical theory of the molecular motions of gases that Maxwell and Boltzmann were to discuss the status of the second law of thermodynamics in relation to the programme of mechanical explanation. Maxwell's first discussion of the issue occurred in a letter written to Tait in 1867, which suggested a way in which a hot body could take heat from a colder body without the performance of external work on the system. Maxwell considered a gas in a vessel divided into two sections, A and B, by a diaphragm. The gas in A was supposed to be hotter than the gas in B, and although Maxwell's velocity distribution law implied that the molecules in both A and B would have a range of velocities of all magnitudes, the higher temperature in A had the consequence that the gas molecules in A had a higher average kinetic energy than the molecules in B. Maxwell then imagined a 'finite being' (later termed a 'demon' by William Thomson) who could observe the motions and velocities of the individual gas molecules. This 'being' would open and close a hole in the diaphragm, and would alternately let molecules from A and from B pass through the hole, selecting the molecules so that faster molecules in B would pass into A, while slower molecules from A would pass into B. The result of this process would be that 'the energy in A is increased and that in B is diminished; that is, the hot system has got hotter and the cold colder and yet no work has been done, only the intelligence of a very observant and neat-fingered being has been employed'.

The purpose of this ingenious argument was not to speculate on the physical possibility of manipulating molecules individually in this way, and hence to violate the second law of thermodynamics, which asserted that heat could not pass from a colder to a warmer body without the performance of external work on the system. Nor was it Maxwell's intention to suggest that a being of a peculiar kind could manipulate molecules in this way. He objected to Thomson's use of the term 'demon', urging Tait to 'call him no more a demon but a valve'; there were no supernatural resonances to the argument. Maxwell's intention was to show that the second law of thermodynamics was a *statistical* law describing the properties of a system of an immense number of molecules, and did not describe the behaviour of an individual molecule within that system. Although a 'finite being' with 'sharpened' faculties would be required to produce an observable flow of heat from a cold body to a hotter one in violation of the second law of thermodynamics, this process occurred spontaneously at the molecular level. Spontaneous fluctuations of individual molecules were continuously occurring in which heat was being transferred from a cold body to a hotter one by the random motions of the molecules. These random fluctuations did not constitute a violation of the second law of thermodynamics, for that law described the observable flow of heat, not the random fluctuations of molecules. As Maxwell observed to Tait, his aim had been 'to show that the 2nd law of thermodynamics has only a statistical certainty'. The 'demon' paradox illustrated the way in which molecular velocities in a gas were distributed statistically and highlighted the existence of spontaneous fluctuations within a system composed of an immense number of molecules; accordingly, it implied the essentially statistical nature of the second law of thermodynamics.

Maxwell published his views on the 'limitation of the second law of thermodynamics' in his *Theory of heat* (1871), and further emphasised there the implications of his argument. He stressed that there was nothing inconsistent with the laws of mechanics and the law of the conservation of energy in supposing a 'mechanism' capable of violating the second law of thermodynamics. Because a 'mechanism' to 'guide and control' the motions of the molecules so as to violate the second law of thermodynamics was consistent with the laws of mechanics, the second law of thermodynamics was not a 'dynamical' law that would describe the motions of individual molecules; hence the physicist was compelled to adopt the 'statistical method of calculation, and to abandon the strict dynamical method'.

The implication of Maxwell's argument was that any molecular interpretation of the second law of thermodynamics must be based on a statistical analysis of the motions of an immense number of molecules; a dynamical explanation of the second law of thermodynamics based on a theory of individual molecular motions was impossible. Though Maxwell made no attempt to formulate a statistical theory of the second law of thermodynamics, he emphasised the implications of his distinction between statistical and dynamical explanations for the conceptual status of thermodynamics. He dismissed as illusory Clausius's attempt to reduce the second law of thermodynamics to a theory of molecular configuration, by means of an explanation of entropy in terms of disgregation (which expressed molecular arrangement). He rejected attempts by Clausius and Boltzmann to provide a dynamical interpretation of the second law of thermodynamics. In 1866 Boltzmann had contrasted the ambiguous conceptual status of the second law of thermodynamics with the secure status of the law of energy conservation. Boltzmann aimed to provide a general proof of the second law of thermodynamics and to discover the 'theorem in mechanics that corresponds to it'. Although he did not succeed in formulating a completely general proof, he derived a dynamical analogue of entropy, a result that Clausius also obtained in 1871, though he expressed it in terms of disgregation. Maxwell regarded this work as fundamentally misconceived: The second law of thermodynamics was an irreducibly statistical law.

By the early 1870s Boltzmann had himself adopted the view that the second law of thermodynamics was a statistical theorem, and could not be derived as a strictly dynamical law. Following his study of Maxwell's theory of gases, Boltzmann published a series of papers in which he developed Maxwell's treatment of the statistical theory of molecular motions, using the velocity distribution law to formulate a statistical proof of the second law of thermodynamics. In a major paper on the thermal equilibrium of gas molecules, published in 1872, he gave a general proof of the uniqueness of Maxwell's distribution law, demonstrating that whatever the initial state of a gas, Maxwell's velocity distribution law would describe its equilibrium state. Boltzmann also derived a formula that expressed the increase of the entropy of an isolated system whenever an irreversible process occurred. This result, later called the 'H theorem', employed the statistical distribution law of molecular motions to establish the concept of the irreversible increase of entropy. Boltzmann therefore used a statistical analysis of molecular motions to establish the second law of thermodynamics.

At this time, however, Boltzmann did not regard the second law of thermodynamics as an essentially statistical law of nature. His formulation of the H theorem asserted that entropy would necessarily increase in irreversible processes; the increase in entropy was expressed as a certainty, not a probability. The problem at issue was the conceptual status of irreversibility, and the relationship between the laws of dynamics and the irreversibility of natural processes that was asserted by the second law of thermodynamics. Boltzmann confronted the problem in response to criticisms raised by Josef Loschmidt (1821–95). Loschmidt's argument, later termed the 'irreversibility paradox', was that irreversibility was a contingent, not a necessary, feature of the natural world. The motion of a system of particles towards an equilibrium state would be accompanied by an increase of entropy; because the equations of motion of the particles were time invariant, the time reversal of these motions from the equilibrium state to a less uniform state would be accompanied by a decrease in entropy. Boltzmann amplified Loschmidt's terse statement of the issue in an attempt to clarify his own views, stressing that the irreversible increase in entropy could not be derived from the laws of mechanics, because the equations of motion of the particles were unaffected by time reversal.

Maxwell had also privately discussed this problem, observing in a letter to J. W. Strutt (later Lord Rayleigh) in 1870 that the time reversal of all events was consistent with the laws of dynamics but inconsistent with the second law of thermodynamics. Maxwell's conclusion was that the irreversibility of natural processes asserted by the second law of thermodynamics could not be explained by dynamical principles. In response to Loschmidt's paradox, Boltzmann adopted a statistical interpretation of entropy and irreversibility. He noted the intimate connection between the second law of thermodynamics and the theory of probability and revised his interpretation of entropy. Conceding that it was possible to imagine processes in which entropy was found to decrease, he argued that this did not vitiate the concept of irreversibility asserted by the second law of thermodynamics. It was not possible to prove that entropy increased with 'absolute necessity' because the increase of entropy was an essentially statistical law. Though entropy-decreasing processes were 'exceedingly improbable', they were nevertheless 'not absolutely impossible'.

In a major paper on the connection between the second law of thermodynamics and the theory of probability, published in 1877, Boltzmann amplified this interpretation of entropy, enunciating the relationship between the entropy of a system and the possible molecular configurations of the system. He defined the second law of thermodynamics as a statistical law, the entropy of a system being a measure of its probability. The increase of entropy in natural processes corresponded to the tendency of systems to reach the most probable molecular distribution. The second law of thermodynamics thus stated that the irreversibility of natural processes was a consequence of the tendency of systems to reach the most probable thermodynamic state, the state of thermal equilibrium. The irreversible increase of entropy in nature was therefore characterised as an irreducibly statistical law.

Boltzmann believed that by using the statistical theory of molecular motions to formulate the concepts of entropy and irreversibility, he had established the second law of thermodynamics within the ontology of the mechanical view of nature. Just as the law of energy conservation was grounded on mechanical principles, in terms of the energy of molecular motions, the second law of thermodynamics was established within the framework of the mechanical view of nature, in terms of the statistical theory of molecular motions.

Chemical Thermodynamics and Energetics

In the 1880s and 1890s Boltzmann's approach to thermodynamics, based on the statistical theory of molecular motions, was challenged. In an 1891 lecture Max Planck (1858–1947) delineated the difficulties encountered in the attempt to relate the statistical theory of molecular motions to thermodynamic concepts. The problems that the equipartition theorem raised for the kinetic theory of gases, and the formidable mathematical complexities of Boltzmann's theory of molecular probabilities, demonstrated the difficulties in elaborating the kinetic theory of gases. Planck believed that the work of Maxwell and Boltzmann, the physicists who had provided the most penetrating analyses of the theory of molecular motions, showed that the achievements of the theory did not compare with the physical and mathematical sophistication employed in elaborating it. Planck questioned the attempt to explain thermodynamic concepts by the statistical theory of molecular motions, challenging Boltzmann's theory of nature. Planck contrasted the ambiguities of Boltzmann's programme with the successes achieved in the work of theorists who had employed the concepts of energy and entropy without reference to a theory of molecular motions. He pointed to the development of chemical thermodynamics, such as the work of Josiah Willard Gibbs (1839–1903) on the application of entropy and energy to the study of chemical processes. In a paper published in 1876 Gibbs had

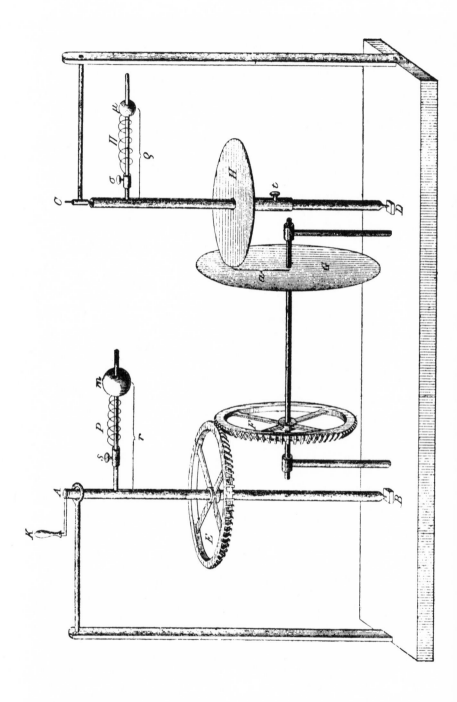

Fig. 5.4. Boltzmann's mechanical model to illustrate the second law of thermodynamics (1884). This model was an illustration of an argument developed by Helmholtz in 1884 to demonstrate the limited convertibility of heat into work as implied by the second law of thermodynamics (which stated that in a cyclic process in which the heat of a body was converted into work, heat would pass from the hot body to a colder one). Helmholtz supposed that molecular motions in a gas were analogous to rotations of a wheel about a fixed axle, and that the energy of the wheel was dependent only on its angular velocity. If mechanical devices were attached to the wheel, the energy of the system would also be dependent on the other coordinates, which could be varied at a slow rate compared to the speed of rotation. Helmholtz established that the energy supplied to the system as heat, represented by a change in the kinetic energy of the wheel (corresponding to molecular motions in a gas), could not be completely converted into work, represented by a slow change in the other parameters of the mechanical system (corresponding to the volume of the gas). Although Helmholtz disavowed any claim to have provided a mechanical explanation of the second law of thermodynamics, he asserted that he had provided a mechanical analogy for thermodynamics based on the equations of a mechanical system.

In illustrating Helmholtz's argument by a mechanical model, Boltzmann explored the implications of Helmholtz's mechanical analogue for thermodynamics. By 1877 Boltzmann had formulated a completely statistical interpretation of the second law of thermodynamics, and in elaborating Helmholtz's mechanical interpretation of thermodynamics Boltzmann did not retreat from his view of this law as irreducibly statistical. He wished to explore a mechanical analogue as a way of analysing problematic features of thermodynamic systems, such as the distinction between heat and work at the molecular level, and he emphasised Helmholtz's distinction between the molecular coordinates of the system (corresponding to heat) and the slowly varying parameters (corresponding to work). *Source:* Ludwig Boltzmann, *Wissenschaftliche Abhandlungen,* ed. F. Hasenöhrl, 3 vols. (Leipzig, 1909), 3:143.

formulated the concept of the thermodynamic equilibrium of a system in terms of energy and entropy, and had applied this concept to the problem of chemical equilibrium. The development of chemical thermodynamics, including work by Helmholtz and Planck himself on the application of the entropy concept to chemical reactions, offered a basis for thermodynamics that was an alternative to Boltzmann's interpretation of entropy.

The problem of reaction mechanisms had slipped from the centre of chemical debate in the nineteenth century. In the eighteenth century the theory of chemical affinities was employed to explain chemical reactions, but although Berthollet transformed affinity theory into a more quantitative chemistry, he also made the study of chemical mechanisms seem impossibly complex. Dalton's chemical atomic theory, with its emphasis on the quantification of particle weights, provided a rationale for the laws of chemical proportions and offered an alternative and seemingly more promising basis for the study of chemical substances. Berthollet's view that compounds formed in indefinite proportions, that chemical particles combined in proportions dictated by the forces of chemical affinity, contradicted Dalton's theory of definite proportions, which rapidly gained acceptance by chemists. Dalton's work fostered the shift in the focus of chemical research from the study of reaction mechanisms to the study of combining proportions and atomic weights.

In the 1850s there was a new interest in Berthollet's ideas. The measurement of heats of reaction stimulated work on reaction mechanisms and chemical equilibria, and the development of the kinetic theory of gases suggested a model of particles in motion, rather than of static attractive forces, for the explanation of chemical reactions, providing a more satisfactory physical model for chemical reactions. The nature of chemical solutions, which have indefinite proportions, became a subject of interest, and Berthollet's theory of affinity stimulated discussion of the problems of chemical equilibria. The study of the thermal effects of chemical combination came to be seen as providing information about chemical reactions, and by the 1880s, in the work of Gibbs, Helmholtz, and Planck, thermodynamic concepts were applied to the study of chemical processes. Reaction mechanisms, the nature of chemical affinity, the theory of chemical equilibria, and the direction of chemical reactions were brought within the framework of thermodynamic explanation.

The emergence of the new discipline of 'physical chemistry' in the 1880s was intimately connected with the development of chemical thermodynamics, and especially with the study of the theory of solutions. Jacobus Henricus van't Hoff (1852–1911) argued that there was an analogy between dilute chemical solutions and gases, and developed a thermodynamic treatment of the chemistry of solutions. Svante Arrhenius (1859–1927) applied the electrochemical concept of the dissociation of certain chemical compounds into charged 'ions' to the theory of solutions, relating the degree of dissociation to dilution. Friedrich Wilhelm Ostwald (1853–1932) enlarged upon these ideas, and these chemists developed a systematic theory of solutions based on the concepts of ionic dissociation and the analogy between gases and chemical solutions, emphasising the application of thermodynamic concepts to the study of chemical processes. This programme of research formed the basis of their physical chemistry.

Ostwald elaborated this thermodynamic programme into a general theory of 'energetics', seeking to demonstrate that the use of atomistic concepts to explain thermodynamics was mistaken. He argued that the future task of physico-chemical science would be the development of the energy concept. Energy was the single real entity in nature, and matter was derivative, a manifestation of the disposition of energy. Ostwald was opposed to the use of atomistic concepts in chemistry, and he rejected all associated concepts, the kinetic theory of gases, and the programme of mechanical explanation in physics. Although Planck criticised Boltzmann's grounding of entropy on the statistical theory of molecular motions, he also rejected Ostwald's view of the status of the energy concept. Ostwald regarded entropy as a quantity representing the dissipation of energy, but Planck declared that the second law of thermodynamics could not be interpreted in terms of the energy concept. In his own writings on thermodynamics, Planck had endeavoured to delineate the conceptual status of entropy, stressing the crucial significance of irreversibility and the importance of entropy for the direction of natural processes, rather than seeking an interpretation of entropy according to a theory of molecular motions. In response to Ostwald's arguments, Planck stressed that any attempt to explain irreversibility must distinguish between the entropy values of the initial and final states of irreversible processes.

Boltzmann and Planck were in agreement in pointing out the failure of Ostwald's 'energetics' to provide an adequate account of the fundamental importance of entropy. While Boltzmann continued to expound his interpretation of entropy and irreversibility as essentially statistical concepts, Planck criticised the probabilistic interpretation of entropy. In his writings on thermodynamics in the 1890s he proferred the view that the second law of thermodynamics possessed absolute certainty, avoiding a statistical interpretation of entropy. But Planck did not dismiss the ontological assumptions of the mechanical view of nature. He pointed out that the kinetic theory of gases and the analysis of molecular motions provided in principle a perfect explanation of thermodynamic processes. The conceptual difficulties of Boltzmann's statistical theory, however, led Planck to prefer a purely thermodynamic, rather than statistical, explanation of entropy; nevertheless, Planck ultimately came to accept Boltzmann's view of entropy. As a result of his study of the irreversibility of radiation processes (which led to his introduction of the quantum theory in 1900), Planck adopted a probabilistic interpretation of entropy, eventually even abandoning his commitment to the absolute validity of the second law of thermodynamics and accepting Boltzmann's completely statistical view of this law.

The relationship between the laws of thermodynamics and the statistical theory of molecular motions remained the subject of debate. Reviewing the problem in his *Elementary principles in statistical mechanics* (1902), Gibbs emphasised the distinction between the formulation of hypotheses about molecular actions and the constitution of material bodies, and the statement of the laws of thermodynamics. He described his treatment of the foundations of thermodynamics, in terms of a statistical theory of molecular motions, as the attempt to formulate thermodynamic analogies. The elaboration of a 'statistical mechanics' of molecular motions brought thermodynamics within the conceptual framework of the mechanical view of nature, by providing molecular analogues for thermodynamic concepts. Remarking the gap between the laws of thermodynamics and their molecular analogues, Gibbs nevertheless affirmed that thermodynamics could be interpreted within the ontology of the mechanical view of nature.

CHAPTER VI

Epilogue: The Decline of the Mechanical World View

In his 1900 lecture 'Nineteenth century clouds over the dynamical theory of heat and light', William Thomson pointed to two problems facing the mechanical theory of nature: the failure to explain the mechanism of the motion of the earth through the ether, and the difficulty the concept of the equipartition of energy posed for the construction of molecular models. Thomson highlighted two 'clouds' that threatened his elaboration of mechanical models of physical phenomena, but there were wider dimensions to the difficulties that physicists perceived in the conceptual rationale of the mechanical theory of nature.

The traditional programme of mechanical explanation elicited diverse responses from physicists in the 1880s and 1890s. Thomson's ether models and Boltzmann's lectures on field theory continued the programme of elaborating detailed mechanical models of phenomena. Boltzmann strove to provide an exhaustive treatment of every detail of the structure and motions of his mechanical models of the electromagnetic field; and Thomson declared that the construction of a mechanical model of a phenomenon was the criterion of the intelligibility of that phenomenon. Nevertheless, the conceptual difficulties associated with the enunciation of mechanical models were well understood. Maxwell had pointed out that such models could not provide unique explanations of phenomena and had drawn attention to the dangers of confusing representation and reality, and though he remained committed to the ultimate aim of formulating a 'complete' mechanical theory of the field, in his *Treatise* he employed an analytical formulation of dynamics, rather than a specific mechanical model. Larmor maintained that this analytical formalism provided a sufficient explanation of the mechanical structure of the electromag-

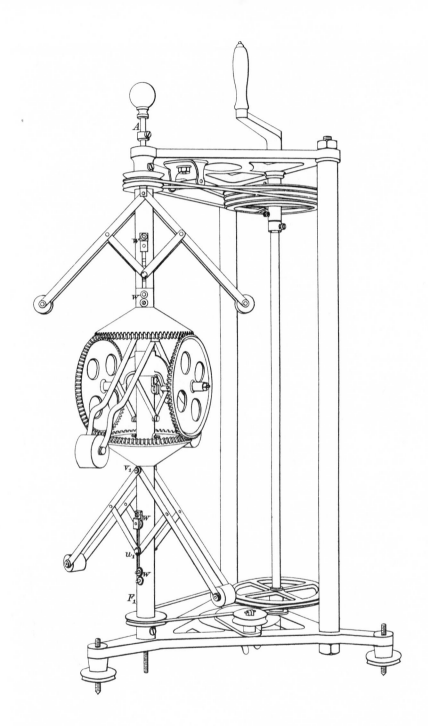

netic field, and he developed a theory of an ethereal plenum to unify the electromagnetic and mechanical properties of the ether. In Larmor's theory electrons were centres of rotational strain in an ether endowed with dynamical properties, and also fundamental electromagnetic entities.

Hertz also affirmed the programme of mechanical explanation, emphasising his commitment to the concept of an ether whose parts were connected by a mechanical structure; but his careful distinction between the formalism of electromagnetism and its representation by a mechanical model fostered the development of the idea that the electromagnetic field possessed no mechanical properties. In his first formulation of his electron theory, Lorentz had conceived the electromagnetic field as a dynamical system, but he had ultimately abandoned reference to the formalism of analytical dynamics in favour of an ontology of electrons and the electromagnetic ether that was not grounded on a framework of mechanical principles. In the electromagnetic world view the mechanical properties of matter were conceived as being grounded on the properties of the electromagnetic ether. Electromagnetic rather than mechanical concepts came to be seen as providing the

---

Fig. 6.1. Mechanical model for induction between electric circuits from Boltzmann's lectures on Maxwell's theory (1891). Boltzmann invented this working mechanical model, which was constructed to his specifications, to represent inductively coupled circuits. The flow of electric currents and the interactions between electric circuits were represented by the rotation of meshing discs, the energy of an electric current being interpreted as dependent on the corresponding velocity of a rotating disc. Boltzmann established the identity of the equations of his mechanical system and those for induction between circuits. He was not satisfied with the analytic representation of Maxwell's theory formulated by Lagrange's dynamical equations, or with a schematic model of the kind postulated by Maxwell in 1861 in his paper 'Physical lines of force'. Boltzmann sought to describe a working mechanical model, explaining its structure and motion in great detail. He argued that mechanical analogies possessed great heuristic value in clarifying the meaning of Maxwell's theory of electricity. Maxwell himself had designed a mechanical model to illustrate the induction of electric currents between two electric circuits; his model, described in Campbell and Garnett's *Life of Maxwell*, was constructed at the Cavendish Laboratory, Cambridge, during his tenure of the Cavendish Professorship of Physics in the 1870s. *Source*: Ludwig Boltzmann, *Vorlesungen über Maxwells Theorie der Elektricität und des Lichtes*, 2 vols. (Leipzig, 1891–3), 1:tab. II, fig. 15.

fundamental and unifying framework of physics, a view echoed in the work of physicists other than Lorentz by around 1900. Lorentz was able to offer an interpretation of the earth's motion through the ether according to his electromagnetic theory of nature, an argument that avoided the difficulties faced by mechanical theories of the elastic solid ether.

Debates about the status of mechanism in field theory had their counterpart in the controversies over thermodynamics. Boltzmann maintained his resolute defence of the intelligibility of the mechanical view of nature, proposing models of molecular structure consistent with the equipartition theorem in an effort to support the conceptual coherence of the kinetic theory of gases, which provided a theory of molecular motions, a paradigm for the mechanical view of nature. By establishing the concepts of entropy and irreversibility by a statistical theory of molecular motions, he tried to interpret the second law of thermodynamics within the ontology of the mechanical view of nature. In questioning Boltzman's attempt to interpret thermodynamic concepts in terms of the statistical analysis of molecular motions, Planck pointed to the difficulties raised by Boltzmann's work. Though Planck did not completely reject the ontology of the mechanical world view, he sought an explanation of entropy in purely thermodynamic terms. The energeticists, however, rejected the ontological assumptions of mechanistic physics along with the supposition of mechanical models, seeking to dislodge all atomistic concepts from the framework of physical theory.

The problems associated with the equipartition theorem and the construction of ether models – Thomson's two 'clouds' over the mechanical theory of nature – heightened the debates over the fundamental structure of physics. In urging the replacement of the mechanical programme by an electromagnetic ontology, Lorentz sought to explain the earth's motion through the ether by appeal to an electromagnetic physics. Although Boltzmann's elaboration of the equipartition theorem sought to lend support to the kinetic theory of gases, and hence to the mechanistic ontology, Boltzmann did not resolve the problem of constructing consistent models of molecular structure; the questionable status of the equipartition theorem threatened the coherence of the kinetic theory of gases. The status of the mechanistic programme as the rationale for physical explanation came to be questioned. Hertz continued to affirm the ontology of the mechanical world view, but his formulation of the equations of field theory as detached from any mechanical interpretation, and Planck's search for a purely thermodynamic interpretation of entropy, fostered the renunciation of the mechanical programme. The decline of mechanical explanation as the paradigm for physical theory was supported by philosophical critiques of mechanical explanation. Ernst Mach (1838–1916) analysed the history of mechanics in support of his contention that the laws of mechanics did not have a privileged status in physics. He questioned the assumption that mechanical explanation was a necessary framework for the comprehensibility of natural phenomena, arguing that the dominant status of mechanics in physical theory was a historical contingency. Mach extended his critique of the mechanistic philosophy of physics to an attack on atomism, rejecting the ontology of the mechanical world view. He emphasised the hypothetical status of the atomic theory, maintaining that atoms were merely symbols for the representation of phenomena, not real physical particles.

Thomson's two 'clouds' were ultimately dispersed, though not by a revision of his mechanistic physics. The quantum theory detached speculation about atomic structure from the constraints imposed by the kinetic theory of gases; and the theory of 'relativity' proposed by Albert Einstein (1879–1955) in 1905 dismissed the need for an ether as superfluous. These developments were shaped by the controversies of the 1890s about the programme of mechanical explanation. Planck's introduction of the quantum of energy in 1900 was the result of his study of the irreversibility of radiation processes. Failing to provide a purely thermodynamic interpretation of radiation processes that would satisfactorily explain his radiation law, he introduced a statistical interpretation of entropy, initiating his ultimate acceptance of Boltzmann's statistical approach to the second law of thermodynamics. Planck's radiation formula also implied that the energy of an oscillator was a discrete variable, rather than a continuously varying quantity. Planck's introduction of the energy quantum occurred in the context of his thermodynamic researches, work that was shaped by the controversies of the 1890s in which he had played a notable part, seeking to define a purely thermodynamic interpretation of entropy.

Einstein consciously directed his work to resolving the controversy between the mechanical world view and Lorentz's electron theory. Einstein's light-quantum hypothesis of 1905 was an attempt to resolve the disjunction between electromagnetism and mechanics, to bridge the 'profound formal distinction' between the formalisms of field theory and particle mechanics, between electromagnetism and the kinetic theory of gases. Rejecting the dualism of discrete electrons and the ether in Lorentz's electron theory,

Einstein sought to bridge the dualism between the electromagnetic and mechanical world views by supposing that light was particulate. His paper on 'relativity', published in the same year, amplified his attempt to formulate a unified physics. The fundamental postulates of the theory were universal principles that applied to both mechanics and electrodynamics. His abandonment of the ether as 'superfluous' highlighted his rejection of the electromagnetic world view; accordingly, his theory was interpreted as a principle of mechanics and as old-fashioned by physicists who believed that electromagnetic concepts provided the basis for a universal physics. Einstein had been influenced by Mach's critique of mechanistic physics, and he did not propose to reduce electrodynamics to mechanics, but to transform electrodynamics and mechanics into a more general and fundamental physical theory. Einstein's statement later in the year of the equivalence of mass and energy, mass being regarded as a form of energy, provided a further unification of physical concepts, a physical basis for the unification of electrodynamics and mechanics. Einstein sought to transform the fundamental concepts of mechanics as the basis of his unified physics, and thus to resolve the conflict between the electromagnetic and mechanical world views. The equivalence of mass and energy conflated the categories of mechanistic physics, providing a conceptual framework for a unified physics.

The abandonment of the doctrines of absolute space and time in Einstein's theory of relativity and of causality and determinism in quantum mechanics has traditionally been regarded as marking a 'revolution' in the history of physics. The quantum mechanics of the 1920s displays a radical break with previous physical theory, but though the term 'revolution' usefully indicates that scientific development cannot be represented as a simple cumulative process involving the accretion of scientific facts, the term is ambiguous in that it implicitly denies the continuity of ideas that is present even in episodes of striking conceptual change. The remarkable theoretical innovations of early twentieth-century physics, the development of relativity and quantum theory in the work of Einstein and Planck, cannot be adequately understood detached from their intellectual context, the world-view debates of the 1890s. The prevailing historical image of the intellectual triumphs of sixteenth- and seventeenth-century mechanics and astronomy as the 'Scientific Revolution', and an implicit but erroneous image of 'classical' physics as monolithic, as forming a unified world view, have shaped the interpretation of the emergence of 'modern' physics as a 'revolutionary' break with classical physics. Although the customary emphasis on the discontinuity between classical and modern physics is appropriate when used to distinguish the philosophical assumptions of eighteenth- and nineteenth-century physics from the relativistic and indeterministic doctrines of twentieth-century physics, and to distinguish physics before and after the development of quantum mechanics in the 1920s, this disjunction is overemphatic, and misleadingly ignores the continuity of ideas between the classical and modern periods.

Bibliographic Essay

Introduction

Many important nineteenth-century physics texts are available in modern reprints, and some nineteenth-century biographies are of value, especially as sources of correspondence and contemporary attitudes. The *Dictionary of scientific biography*, ed. C. C. Gillispie, 16 vols. (New York, 1970–80), provides articles on most of the important nineteenth-century physical scientists. Some of these articles are substantial studies, and many include good bibliographies. A classic study of the development of physics is provided by J. T. Merz in his *A history of European thought in the nineteenth century*, 4 vols. (reprint ed., New York, 1965), 1:302–458, 2:3–199, a valuable and wide-ranging survey. The treatment of the philosophical structure of physical theory by E. Meyerson, *Identity and reality*, trans. K. Loewenberg (London, 1930), is a notable study of classical physics. M. Capek, *The philosophical impact of contemporary physics* (Princeton, 1961), provides an analysis of the conceptual foundations of classical physics. J. B. Stallo, *The concepts and theories of modern physics* (reprint ed., Cambridge, Mass., 1960), gives a contemporary discussion of nineteenth-century physics, written from a strongly critical standpoint.

Nineteenth-century physics is discussed by P. M. Heimann [Harman] in 'The scientific revolutions', *The new Cambridge modern history: XIII. companion volume*, ed. P. Burke (Cambridge, 1979), 248–70. There is a general discussion of the development of the physics discipline by T. S. Kuhn, 'Mathematical versus experimental traditions in the development of physical science', *Journal of Interdisciplinary History* 7 (1976), 1–31, reprinted in Kuhn's *The essential tension* (Chicago, 1977), pp. 31–65. R. McCormmach's

156

'Introduction' to *Historical Studies in the Physical Sciences* 3 (1971) gives a useful perspective on this topic. C. C. Gillispie, *The edge of objectivity* (Princeton, 1960), pp. 352–520, surveys nineteenth-century physics.

Though there are no systematic studies of the institutionalisation of physics in the nineteenth century, there is a valuable account of the institutional setting of physics in the early modern period by J. L. Heilbron, *Electricity in the seventeenth and eighteenth centuries: a study of early modern physics* (London, 1979), pp. 98–166. The monograph by P. Forman, J. L. Heilbron, and S. Weart, 'Physics *circa* 1900: personnel, funding, and productivity of the academic establishments', *Historical Studies in the Physical Sciences* 5 (1975), establishes a bibliographic and taxonomic basis for the organisation of physics in the nineteenth century. Two papers by R. Sviedrys, 'The rise of physical science at Victorian Cambridge', *Historical Studies in the Physical Sciences* 2 (1970), 127–45, and 'The rise of physics laboratories in Britain', ibid. 7 (1976), 405–36, deal with British academic physics. On physical science in Germany there are useful discussions by R. S. Turner, 'The growth of professorial research in Prussia, 1818 to 1848: causes and context', *Historical Studies in the Physical Sciences* 3 (1971), 137–82, and C. Jungnickel, 'Teaching and research in the physical sciences and mathematics in Saxony, 1820–1850', ibid. 10 (1979), 3–47. R. Fox, 'Scientific enterprise and the patronage of research in France, 1800–70', *Minerva* 11 (1973), 442–73, and T. Shinn, 'The French science faculty system: institutional change and research potential in mathematics and the physical sciences', *Historical Studies in the Physical Sciences* 10 (1979), 271–332, give emphasis to the status of the physical sciences in France. J. Cawood, 'Terrestrial magnetism and the development of international collaboration in the early nineteenth century', *Annals of Science* 34 (1977), 551–87, discusses the growth of the physics community.

The Context of Physical Theory: Energy, Force, and Matter

C. A. Truesdell, *Essays in the history of mechanics* (Berlin, 1968), is a valuable introduction to rational mechanics. The systematic study by J. L. Heilbron, *Electricity in the seventeenth and eighteenth centuries* (London, 1979), describes the development of the science of electricity to its transformation into the quantified physics of the early nineteenth century, and analyses the structure of physical theory in the early modern period. Essays by H. J. M. Bos, 'Mathematics and rational mechanics', and by J. L. Heilbron,

'Experimental natural philosophy', in *The ferment of knowledge: studies in the historiography of eighteenth-century science*, ed. G. S. Rousseau and R. Porter (Cambridge, 1980), pp. 327–55 and 357–87, provide useful perspectives on eighteenth-century physics. P. M. Heimann [Harman], 'Ether and imponderables', in *Conceptions of ether: studies in the history of ether theories, 1740–1900*, ed. G. N. Cantor and M. J. S. Hodge (Cambridge, 1981), pp. 61–83, traces the development of imponderable fluid theories. A. W. Thackray, *Atoms and powers: an essay on Newtonian matter-theory and the development of chemistry* (Cambridge, Mass., 1970), discusses the Newtonian tradition in eighteenth-century chemistry. On the caloric theory of heat, H. E. Guerlac, 'Chemistry as a branch of physics: Laplace's collaboration with Lavoisier', *Historical Studies in the Physical Sciences* 7 (1976), 193–276, analyses an influential work. R. Fox, *The caloric theory of gases from Lavoisier to Regnault* (Oxford, 1971), gives a detailed account of the most sophisticated and successful dimension of the caloric theory.

The role of Laplace and his associates in shaping the structure of early nineteenth-century physics has received considerable attention. M. P. Crosland, *The society of Arcueil: a view of French science at the time of Napoleon I* (London, 1967), surveys the work of the Laplace–Berthollet school. R. Fox, 'The rise and fall of Laplacian physics', *Historical Studies in the Physical Sciences* 4 (1974), 89–136, and E. Frankel, 'J. B. Biot and the mathematization of experimental physics in Napoleonic France', ibid. 8 (1977), 33–72, discuss Laplacian quantitative physics. E. Frankel's paper 'The search for a corpuscular theory of double refraction: Malus, Laplace and the prize competition of 1808', *Centaurus* 18 (1974), 223–45, provides a detailed analysis of Laplacian optics. R. Fox 'The background to the discovery of Dulong and Petit's law', *British Journal for the History of Science* 4 (1968), 1–22, discusses the changing fortunes of Laplacian heat theory. M. P. Crosland, *Gay-Lussac: scientist and bourgeois* (Cambridge, 1979), is a biography of a leading member of the Laplace–Berthollet group. There is an account by M. Crosland and C. W. Smith entitled 'The transmission of physics from France to Britain: 1800–1840', *Historical Studies in the Physical Sciences* 9 (1978), 1–61.

A useful discussion of early nineteenth-century British physical science is R. E. Schofield's *Mechanism and materialism: British natural philosophy in an age of reason* (Princeton, 1970), pp. 277–97. The biography of Young by A. Wood, *Thomas Young, natural philosopher* (Cambridge, 1954), surveys Young's varied intellectual activities. Young's *Course of lectures on natural philosophy and the mechanical arts*, 2 vols. (London, 1807), collects his optical papers and is a valuable source on physical theory circa 1800. There is a biography of Rumford by S. C. Brown, *Benjamin Thompson, Count Rumford* (London, 1979). There are several discussions of theories of heat, light, and electricity in early nineteenth-century Britain: R. Olson, 'Count Rumford, Sir John Leslie, and the study of the nature and propagation of heat at the beginning of the nineteenth century', *Annals of Science* 26 (1970), 273–304; S. J. Goldfarb, 'Rumford's theory of heat: a re-assessment', *British Journal for the History of Science* 10 (1977), 25–36; G. N. Cantor, 'The changing role of Young's ether', ibid. 5 (1970), 44–62; Cantor, 'Henry Brougham and the Scottish methodological tradition', *Studies in History and Philosophy of Science* 2 (1971), 69–89, which discusses the reaction to Young's papers; and T. H. Levere, *Affinity and matter: elements of chemical philosophy, 1800–1865* (Oxford, 1971), largely concerned with Davy and Faraday. C. W. Smith, ' "Mechanical philosophy" and the emergence of physics in Britain: 1800–1850', *Annals of Science* 33 (1976), 3–29, reviews the transformation in British physics.

On Fresnel and the optical ether, E. T. Whittaker, *A history of the theories of aether and electricity: I. the classical theories* (London, 1951), is a classic study. There is a useful survey by K. F. Schaffner, *Nineteenth-century aether theories* (Oxford, 1972), pp. 40–75, including extracts of texts by Fresnel, Green, Stokes, MacCullagh, and later theorists. Fresnel's papers on diffraction are collected in the first volume of his *Oeuvres complètes*, ed. H. de Senarmot, E. Verdet, and L. Fresnel, 3 vols. (Paris, 1866–70). R. H. Silliman, 'Fresnel and the emergence of physics as a discipline', *Historical Studies in the Physical Sciences* 4 (1974), 137–62, discusses the wider implications of Fresnel's work. E. Frankel, 'Corpuscular optics and the wave theory of light: the science and politics of a revolution in physics', *Social Studies of Science* 6 (1976), 141–84, is a detailed study of the impact of Fresnel's optics. J. Z. Buchwald, 'Optics and the punctiform ether', *Archive for History of Exact Sciences* 21 (1980), 245–78, and 'The quantitative ether in the first half of the nineteenth century', in *Conceptions of ether*, ed. G. N. Cantor and M. J. S. Hodge (Cambridge, 1981), pp. 215–37, review the molecular ether models of Fresnel, Cauchy, and other theorists. G. N. Cantor, 'The reception of the wave theory of light in Britain', *Historical Studies in the Physical Sciences* 6 (1975), 109–32, analyses the methodological issues in the discussion of Fresnel's work by British physicists. The papers of Green, MacCullagh, and Stokes are collected in the *Mathematical papers of the late George Green*, ed.

N. M. Ferrers (Cambridge, 1871); *The collected works of James MacCullagh*, ed. J. H. Jellett and S. Haughton (Dublin, 1880); and Stokes's *Mathematical and physical papers*, 5 vols. (Cambridge, 1880–1905).

The development of mathematical physics in France after 1815, especially the work of Fourier, has attracted some attention. Fourier's *Analytical theory of heat*, trans. A. Freeman (reprint ed., New York, 1955), is a fundamental source for the history of physics in the nineteenth century. The article by J. R. Ravetz and I. Grattan-Guinness entitled 'Fourier' in the *Dictionary of scientific biography*, 5:165–71, provides a succinct account of Fourier's achievement. There is a major study by I. Grattan-Guinness, in collaboration with J. R. Ravetz, *Joseph Fourier, 1768–1830: a survey of his life and work, based on a critical edition of his monograph on the Propagation of Heat, presented to the Institut de France in 1807* (Cambridge, Mass., 1970), that provides a substantial commentary on Fourier's mathematical physics. J. W. Herivel, *Joseph Fourier: the man and the physicist* (Oxford, 1975), joins a biography to a study of Fourier's physics; there is a detailed review of this book by I. Grattan-Guinness in *Annals of Science* 32 (1975), 503–14. An essay by R. M. Friedman, 'The creation of a new science: Joseph Fourier's analytical theory of heat', *Historical Studies in the Physical Sciences* 8 (1977), 73–99, analyses Fourier's physical theory and methodology; and I. Grattan-Guinness's 'Joseph Fourier and the revolution in mathematical physics', *Journal of the Institute of Mathematics and Its Applications* 5 (1969), 230–53, discusses Fourier's mathematical methods. J. W. Herivel, 'Aspects of French theoretical physics in the nineteenth century', *British Journal for the History of Science* 3 (1966), 109–32, is concerned with the methodology of French physics.

S. G. Brush, 'The wave theory of heat: a forgotten stage in the transition from the caloric theory to thermodynamics', *British Journal for the History of Science* 5 (1970), 145–67, is an amply documented study, tracing the creation of a theory of heat analogous to the wave theory of light. J. L. Heilbron, 'The electrical field before Faraday', in *Conceptions of ether*, ed. G. N. Cantor and M. J. S. Hodge (Cambridge, 1981), pp. 187–213, is a detailed analysis of ether theories of electricity in the early nineteenth century, especially the work of Oersted and Ampère. K. L. Caneva, 'Ampère, the etherians, and the Oersted connection', *British Journal for the History of Science* 13 (1980), 121–38, provides further discussion. R. C. Stauffer, 'Speculation and experiment in the background of Oersted's discovery of electromagnetism', *Isis* 48

(1957), 33–50, discusses the role of *Naturphilosophie* in shaping Oersted's work. The influence of *Naturphilosophie* on physics has provoked some discussion. B. Gower, 'Speculation in physics: the history and practice of *Naturphilosophie*', *Studies in History and Philosophy of Science* 3 (1973), 301–56, provides a valuable account of the ideas of F. W. J. von Schelling and J. W. Ritter, and a refutation of some of the claims for their influence. There is an important discussion by K. L. Caneva, 'Conceptual and generational change in German physics: the case of electricity, 1800–1846' (diss., Princeton Univ. 1974), pp. 99–103, 132–57, 364–413. This whole subject requires systematic appraisal.

There are many historical studies of the development of the principle of the conservation of energy. A monograph by E. N. Hiebert, *Historical roots of the principle of conservation of energy* (Madison, Wis., 1962), is a valuable study of the context of eighteenth-century mechanics; the bibliography is a useful guide to the older historical literature. W. L. Scott, *The conflict between atomism and conservation theory, 1644–1860* (London, 1970), traces the relation between the theory of hard-body collisions and doctrines of conservation. T. L. Hankins, 'Eighteenth-century attempts to resolve the *vis viva* controversy', *Isis* 56 (1965), 281–97, and P. M. Heimann, ' "Geometry and nature": Leibniz and Johann Bernoulli's theory of motion', *Centaurus* 21 (1977), 1–26, discuss important writings on the conservation of 'living force'. D. S. L. Cardwell, 'Some factors in the early development of the concepts of power, work and energy', *British Journal for the History of Science* 3 (1967), 209–24, discusses the clarification of the concept of work in the early nineteenth century.

The emergence of the law of energy conservation in the 1840s is analysed in a major essay by T. S. Kuhn, 'Energy conservation as an example of simultaneous discovery', in *Critical problems in the history of science*, ed. M. Clagett (Madison, Wis., 1959), pp. 321–56, reprinted in Kuhn, *The essential tension* (Chicago, 1977), pp. 66–104; this essay is an attempt to delineate factors close to the surface of scientific consciousness that were required for the full statement of energy conservation. There is an account by Y. Elkana, *The discovery of the conservation of energy* (London, 1974); and there is a review of Elkana's book by P. Clark, *British Journal for the Philosophy of Science* 27 (1976), 165–76. J. Forrester, 'Chemistry and the conservation of energy: the work of James Prescott Joule', *Studies in History and Philosophy of Science* 6 (1975), 273–313, provides an analysis of the changing focus of Joule's interests and includes a bibliography of writings on Joule. E.

Mendoza, 'The surprising history of the kinetic theory of gases', *Memoirs and Proceedings of the Manchester Literary and Philosophical Society* 105 (1962–3), 15–28, and H. J. Steffens, *James Prescott Joule and the concept of energy* (New York, 1979), provide further discussion. Joule's papers were collected in *The scientific papers of James Prescott Joule*, 2 vols. (London, 1884–7; reprint ed., London, 1963).

The doctrine of the 'conversion of forces' and its relation to the energy concept are discussed by P. M. Heimann, 'Conversion of forces and the conservation of energy', *Centaurus* 18 (1974), 147–61. For further discussion see G. N. Cantor, 'William Robert Grove, the correlation of forces, and the conservation of energy', *Centaurus* 19 (1976), 273–90; C. W. Smith, 'Faraday as a referee of Joule's Royal Society paper "On the Mechanical Equivalent of Heat"', *Isis* 67 (1976), 444–9; and D. C. Gooding, 'Metaphysics versus measurement: the conversion and conservation of force in Faraday's physics', *Annals of Science* 37 (1980), 1–29. The work of Mayer and Colding is relevant to the development of energy physics. Mayer's papers are available in a modernised translation: R. B. Lindsay, *Julius Robert Mayer, prophet of energy* (Oxford, 1973). P. M. Heimann, 'Mayer's concept of "force": the axis of a new science of physics', *Historical Studies in the Physical Sciences* 7 (1976), 277–96, gives an analysis of Mayer's physics. Colding's papers have been translated and provided with an introductory essay by P. F. Dahl in *Ludvig Colding and the conservation of energy* (New York, 1972).

Helmholtz's seminal paper 'On the conservation of force' was translated in *Scientific memoirs, natural philosophy*, ed. J. Tyndall and W. Francis (London, 1853; reprint ed., New York, 1966), pp. 114–62; a revised translation has been published by R. Kahl, *Selected writings of Hermann von Helmholtz* (Middletown, Conn., 1971), pp. 3–55. The original memoir, *Ueber die Erhaltung der Kraft* (Berlin, 1847), has been reprinted (Brussels, 1966). Most of Helmholtz's writings on energy physics are collected in the first volume of his *Wissenschaftliche Abhandlungen*, 3 vols. (Leipzig, 1882–95). R. S. Turner, 'Helmholtz', *Dictionary of scientific biography*, 6:241–53, provides a succinct account of Helmholtz's scientific work. There is a biography by L. Koenigsberger, *Hermann von Helmholtz*, trans. and abridged by F. A. Welby (Oxford, 1906). An essay by P. M. Heimann, 'Helmholtz and Kant: the metaphysical foundations of *Ueber die Erhaltung der Kraft*', *Studies in History and Philosophy of Science* 5 (1974), 205–38, analyses the conceptual foundations of Helmholtz's energy physics. Y. Elkana, 'Helmholtz's "Kraft": an illustration of concepts in flux', *Historical Studies in the Physical Sciences* 2 (1970), 263–98, argues that Helmholtz's ideas are ambiguous. The edition by R. S. Cohen and Y. Elkana of Hermann von Helmholtz, *Epistemological writings*, trans. M. F. Lowe, Synthese Library, vol. 79 (Dordrecht, 1977), includes substantial bibliographies. F. L. Holmes's introduction to Justus Liebig, *Animal chemistry* (reprint ed., New York, 1964), provides a detailed study of the physiological context of Helmholtz's work. T. O. Lipman, 'Vitalism and reductionism in Liebig's physiological thought', *Isis* 58 (1967), 167–85, discusses Liebig's 'vitalism' in relation to his concept of the conversion of 'forces'. Essays by P. F. Cranefield, 'The organic physics of 1847 and the biophysics of today', *Journal of the History of Medicine* 12 (1957), 407–23, and C. A. Culotta, 'German biophysics, objective knowledge and Romanticism', *Historical Studies in the Physical Sciences* 4 (1974), 3–38, provide contrasting perspectives on Helmholtz's physiological theories. R. S. Turner, 'The Ohm–Seebeck dispute, Hermann von Helmholtz, and the origins of physiological acoustics', *British Journal for the History of Science* 10 (1977), 1–24, is also relevant.

Energy Physics and Mechanical Explanation

The development of the energy concept was intertwined with the study of thermal processes. D. S. L. Cardwell, *From Watt to Clausius: the rise of thermodynamics in the early industrial age* (London, 1971), provides an account of the early history of thermodynamics, with an emphasis on the technological context. R. Fox has provided a critical edition of S. Carnot, *Réflexions sur la puissance motrice du feu* (Paris, 1978), with an introductory essay and bibliography. There is a translation of Carnot's memoir, together with Clapeyron's version of Carnot's theory and Clausius's first paper on thermodynamics, in the collection edited by E. Mendoza, *Reflections on the motive power of fire by Sadi Carnot and other papers on the second law of thermodynamics by E. Clapeyron and R. Clausius* (New York, 1960). Useful studies on Carnot include E. Mendoza, 'Contributions to the study of Sadi Carnot and his work', *Archives Internationales d'Histoire des Sciences* 12 (1959), 377–96; T. S. Kuhn, 'The caloric theory of adiabatic compression', *Isis* 49 (1958), 132–40, and 'Sadi Carnot and the Cagnard engine', ibid. 52 (1961), 567–74; D. S. L. Cardwell, 'Power technologies and the advance of science, 1700–1825', *Technology and Culture* 4 (1965), 188–207; R. Fox, 'Watt's expansive principle in the work of Sadi Carnot and Nicolas Clément', *Notes and Records of the Royal Society*

24 (1969), 233–53; P. Lervig, 'On the structure of Carnot's theory of heat', *Archive for History of Exact Sciences* 9 (1972), 222–39; and M. J. Klein, 'Carnot's contributions to thermodynamics', *Physics Today* 27 (1974), 23–28. The relation between Lazare and Sadi Carnot is discussed by C. C. Gillispie, *Lazare Carnot, savant* (Princeton, 1971). The collection *Sadi Carnot et l'essor de la thermodynamique* (Paris, 1976) includes essays by Gillispie, Klein, Mendoza, Fox, and Lervig; the paper in that collection by M. J. Klein, 'Closing the Carnot cycle', pp. 213–19, clarifies Clapeyron's version of the Carnot cycle.

The standard biography of William Thomson, by S. P. Thompson, *The life of William Thomson, Baron Kelvin of Largs*, 2 vols. (London, 1910), remains a valuable study and source of documentary material. The obituary notice by J. Larmor, 'William Thomson, Baron Kelvin of Largs, 1824–1907', *Proceedings of the Royal Society*, ser. A 81 (1908), ii–lxxvi, is an informative contemporary assessment of Thomson's work. Thomson's major papers on thermodynamics are collected in the first volume of his *Mathematical and physical papers*, 6 vols. (Cambridge, 1882–1911). The intellectual framework of Thomson's thermodynamics is analysed by C. W. Smith, 'Natural philosophy and thermodynamics: William Thomson and "The Dynamical Theory of Heat" ', *British Journal for the History of Science* 9 (1976), 293–319. Smith gives a detailed account of the interactions among James Thomson, William Thomson, Joule, and Rankine in his study 'William Thomson and the creation of thermodynamics: 1840–1855', *Archive for History of Exact Sciences* 16 (1976), 231–88. There is further discussion by K. Hutchison, 'Mayer's hypothesis: a study of the early years of thermodynamics', *Centaurus* 20 (1976), 279–304. An essay by M. N. Wise, 'William Thomson's mathematical route to energy conservation: a case study of the role of mathematics in concept formation', *Historical Studies in the Physical Sciences* 10 (1979), 49–83, provides an important perspective on Thomson's thermodynamics.

Clausius's major papers on thermodynamics were translated in the nineteenth century in R. Clausius, *The mechanical theory of heat*, ed. T. A. Hirst (London, 1867). The obituary notice by J. W. Gibbs, 'Rudolf Julius Emmanuel Clausius', *Proceedings of the American Academy of Arts and Sciences* 16 (1889), 458–65, is an informative contemporary assessment. An essay by M. J. Klein, 'Gibbs on Clausius', *Historical Studies in the Physical Sciences* 1 (1969), 127–49, provides a penetrating analysis of the history of thermodynamics. There is an important discussion of the entropy concept by S. G. Brush, 'Randomness and irreversibility', *Archive for History of*

Exact Sciences 12 (1974), 1–88. A volume entitled *Miscellaneous scientific papers of W. J. Macquorn Rankine*, ed. W. J. Millar (London, 1881), collects most of Rankine's writings on energy physics. K. Hutchison, 'W. J. M. Rankine and the rise of thermodynamics', *British Journal for the History of Science* 14 (1981), 1–26, gives a perspective on Rankine's work. E. E. Daub, 'Atomism and thermodynamics', *Isis* 58 (1967), 293–303, contrasts the molecular hypotheses of Rankine and Clausius; and Daub's 'Entropy and dissipation', *Historical Studies in the Physical Sciences* 2 (1970), 321–54, discusses the Tait–Clausius controversy. Some documentary materials on the Tait–Tyndall controversy have been published by J. T. Lloyd, 'Background to the Joule–Mayer controversy', *Notes and Records of the Royal Society* 25 (1970), 211–25. Materials on the conversion of 'forces' and Mayer's theory of 'force' are listed in the preceding section.

The emergence of energy physics in the 1850s and 1860s was a major development. D. F. Moyer, 'Energy, dynamics, hidden machinery: Rankine, Thomson and Tait, Maxwell', *Studies in History and Philosophy of Science* 8 (1977), 251–68, discusses Thomson and Tait's dynamics; and C. W. Smith, 'A new chart for British natural philosophy: the development of energy physics in the nineteenth century', *History of Science* 16 (1978), 231–79, focuses on the intellectual framework of Thomson's energy physics. The text by James Clerk Maxwell, *Matter and motion* (London, 1877; reprint ed., New York, n.d.), is a useful introduction to nineteenth-century dynamical theory. The revised (1912) edition of Thomson and Tait's *Treatise on natural philosophy* (Oxford, 1867) has been reprinted (New York, 1962).

On thermodynamics and cosmogeny, J. D. Burchfield, *Lord Kelvin and the age of the earth* (New York, 1975), discusses Thomson's impact on geological debate; and S. G. Brush, *The temperature of history: phases of science and culture in the nineteenth century* (New York, 1978), provides a broad survey of the cultural ramifications of thermodynamics. On theological issues, E. N. Hiebert, 'The uses and abuses of thermodynamics in religion', *Daedalus* 95 (1966), 1046–80, surveys some later developments; and P. M. Heimann, 'The *Unseen Universe*: physics and the philosophy of nature in Victorian Britain', *British Journal for the History of Science* 6 (1972), 73–9, discusses the apologetics of Stewart and Tait. D. B. Wilson, 'Concepts of physical nature: John Herschel to Karl Pearson', in *Nature and the Victorian imagination*, ed. U. C. Knoepflmacher and G. B. Tennyson (Berkeley, 1978), pp. 201–15, reviews the theological arguments of Victorian physicists. The subject requires a systematic study tracing its relation to other traditions of nineteenth-century naturalism.

Matter and Force: Ether and Field Theories

There are several general studies of the origins and development of field theory, offering different perspectives. B. G. Doran, 'Origins and consolidation of field theory in nineteenth-century Britain: from the mechanical to the electromagnetic view of nature', *Historical Studies in the Physical Sciences* 6 (1975), 133–260, is a wide-ranging and detailed analysis, placing emphasis on William Thomson and Larmor and attempting to reappraise the mechanical foundations of nineteenth-century field and ether theories. L. Rosenfeld, 'The velocity of light and the evolution of electrodynamics', *Nuovo Cimento*, supp. 4 (1956), pp. 1630–69, is valuable for its emphasis on Lorenz as well as on Maxwell and Hertz. By contrast, L. P. Williams, *The origins of field theory* (New York, 1966), focuses on Faraday and suggests the impact of Boscovich and *Naturphilosophie* in shaping the origins of the field concept. W. Berkson, *Fields of force: the development of a world view from Faraday to Einstein* (London, 1974), provides an introduction to the ideas of some of the major figures, again with an emphasis on Faraday. J. E. McGuire, 'Forces, powers, aethers and fields', in *Boston studies in the philosophy of science XIV* (Dordrecht, 1974), pp. 119–59, suggests the roots of the field concept in discussions of matter theory. M. B. Hesse, *Forces and fields: the concept of action at a distance in the history of physics* (London, 1961), pp. 189–225, discusses the role of models in field theory. M. Jammer, *Concepts of force: a study in the foundations of dynamics* (Cambridge, Mass., 1957), analyses the status of ideas of force in the history of physics. E. T. Whittaker, *A history of theories of aether and electricity: I. the classical theories* (London, 1951), is a classic study of field and ether concepts. K. Schaffner, *Nineteenth-century aether theories* (Oxford, 1972), pp. 76–117, provides a useful survey of field and ether theories, including a selection of texts by Fitzgerald, Larmor, and Lorentz. The editors' introduction to *Conceptions of ether*, ed. G. N. Cantor and M. J. S. Hodge (Cambridge, 1981), pp. 1–60, reviews ether and field theories.

Faraday's major papers on electromagnetism were collected in his *Experimental researches in electricity*, 3 vols. (London, 1839–55; reprint ed., New York, 1965). *Faraday's diary, being the various philosophical notes of experimental investigation made by Michael Faraday*, ed. T. Martin, 7 vols. (London, 1932–6), contains further material. H. Bence Jones, *The life and letters of Faraday*, 2 vols. (London, 1870), includes correspondence and other documents. *The selected correspondence of Michael Faraday*, ed. L. P. Williams, 2 vols. (Cambridge, 1971), contains much previously unpublished material, but reference to nineteenth-century sources is still necessary.

A basic source on Faraday is the bibliography by A. E. Jeffreys, *Michael Faraday: a list of his lectures and published writings* (London, 1961). There is a general account of Faraday by J. Agassi, *Faraday as a natural philosopher* (Chicago, 1971), and a detailed study by L. P. Williams, *Michael Faraday: a biography* (London, 1965), providing a systematic reconstruction of Faraday's work. Williams's emphasis on the role of Boscovich in shaping Faraday's conception of nature has been questioned by J. B. Spencer, 'Boscovich's theory and its relation to Faraday's researches: an analytic approach', *Archive for History of Exact Sciences* 4 (1967), 184–202. P. M. Heimann, 'Faraday's theories of matter and electricity', *British Journal for the History of Science* 5 (1971), 235–57, documents the influence of discussions of matter theory in eighteenth-century British natural philosophy in shaping the development of Faraday's field theories. This philosophical tradition is analysed in detail by P. M. Heimann and J. E. McGuire in 'Newtonian forces and Lockean powers: concepts of matter in eighteenth-century thought', *Historical Studies in the Physical Sciences* 3 (1971), 233–306. T. H. Levere has published an interesting manuscript in 'Faraday, matter and natural theology: reflections on an unpublished manuscript', *British Journal for the History of Science* 4 (1968), 95–107. D. C. Gooding, 'Conceptual and experimental bases of Faraday's denial of electrostatic action at a distance', *Studies in History and Philosophy of Science* 9 (1978), 117–49, provides further clarification of Faraday's discussions of matter theory; and Gooding's 'Faraday, Thomson, and the concept of the magnetic field', *British Journal for the History of Science* 13 (1980), 91–120, and his 'Final steps to the field theory: Faraday's study of magnetic phenomena, 1845–1850', *Historical Studies in the Physical Sciences* 11 (1981), 231–75, review Faraday's field theory. J. B. Spencer, 'On the varieties of nineteenth-century magneto-optical discovery', *Isis* 61 (1970), 34–51, reviews Faraday's discovery of magneto-optic rotation.

William Thomson's early work on electricity and magnetism is collected in his *Reprint of papers on electrostatics and magnetism* (London, 1872). J. Z. Buchwald, 'William Thomson and the mathematisation of Faraday's electrostatics', *Historical Studies in the Physical Sciences* 8 (1977), 101–36, examines Thomson's early papers. R. H. Silliman, 'William Thomson: smoke rings and nineteenth century atomism', *Isis* 54 (1963), 461–74, discusses the vortex atom theory. O. Knudsen, 'From Lord Kelvin's notebook: ether speculations', *Centaurus* 16 (1972), 41–53, makes available an important manuscript, with a commentary. C. W. Smith, 'Engineering the universe: William Thomson and Fleeming Jenkin on the nature of matter', *Annals of Science* 37 (1980), 387–412, provides a general perspective on Thomson's physical theories.

Maxwell's major works on electromagnetism are collected in *The scientific papers of James Clerk Maxwell*, ed. W. D. Niven, 2 vols. (Cambridge, 1890; reprint ed., New York, 1965), and in Maxwell's *Treatise on electricity and magnetism*, 2 vols. (Oxford, 1873; reprint of 3d ed., New York, 1954). Substantial portions of Maxwell's correspondence were published in his biography and in the biographies of his major contemporaries: L. Campbell and W. Garnett, *The life of James Clerk Maxwell* (London, 1882), with the second edition of 1884 containing additional correspondence included in the reprint of the first edition (New York, 1969); C. G. Knott, *Life and scientific work of Peter Guthrie Tait* (Cambridge, 1911); and *Memoir and scientific correspondence of the late Sir George Gabriel Stokes, Bart.*, ed. J. Larmor, 2 vols. (Cambridge, 1907). Larmor collected Maxwell's important letters to Thomson on electricity in his 'The origins of Clerk Maxwell's electric ideas, as described in familiar letters to William Thomson', *Proceedings of the Cambridge Philosophical Society* 32 (1936), 695–750, reprinted as a separate volume (Cambridge, 1937). Maxwell's introductory lecture at Aberdeen is printed by R. V. Jones in 'James Clerk Maxwell at Aberdeen, 1856–1860', *Notes and Records of the Royal Society* 28 (1973), 57–81.

There is a useful survey of Maxwell's scientific work by C. W. F. Everitt, *James Clerk Maxwell: physicist and natural philosopher* (New York, 1975), based on Everitt's article 'Maxwell', *Dictionary of scientific biography*, 9:198–230. M. N. Wise's articles 'The flow analogy to electricity and magnetism: Part I. William Thomson's reformulation of action at a distance' and 'Part II. Maxwell's first formulation of field theory', *Archive for History of Exact Sciences* 24 (1981), in press, provide a detailed analysis of Thomson's and Maxwell's early work on field theory. D. M. Siegel, 'Thomson, Maxwell and the universal ether in Victorian physics', in *Conceptions of ether*, ed. G. N. Cantor and M. J. S. Hodge (Cambridge, 1981), pp. 239–68, reviews the development of Maxwell's field theory and critically reappraises the concept of the displacement current. P. M. Heimann, 'Maxwell and the modes of consistent representation', *Archive for History of Exact Sciences* 6 (1970), 171–213, provides an account of Maxwell's development of Faraday's ideas and an analysis of Maxwell's physical world view. T. K. Simpson, 'Some observations on Maxwell's *Treatise on electricity and magnetism*', *Studies in History and Philosophy of Science* 1 (1970), 249–63, and D. F. Moyer, 'Continuum mechanics and field theory: Thomson and Maxwell', ibid. 9 (1978), 35–50, clarify Maxwell's concept of dynamical explanation. M. N. Wise 'The mutual embrace of electricity and magnetism', *Science* 203 (1979), 1310–18, analyses the conceptual structure of Maxwell's electromagnetic theory.

Maxwell's concept of electric charge and the displacement current has given rise to much discussion, the classic work, from an adversary standpoint, being P. Duhem's *Les théories électriques de J. Clerk Maxwell: étude historique et critique* (Paris, 1902). J. Bromberg has provided the most detailed recent discussion of the issue: 'Maxwell's displacement current and his theory of light', *Archive for History of Exact Sciences* 4 (1967), 218–34, and 'Maxwell's electrostatics', *American Journal of Physics* 36 (1968), 142–51. The intelligibility of Maxwell's arguments is still under debate. O. Knudsen, 'The Faraday effect and physical theory, 1845–1873', *Archive for History of Exact Sciences* 15 (1976), 235–81, gives a detailed analysis of interpretations of the magneto-optic effect by Maxwell and other physicists. M. J. Crowe, *A history of vector analysis* (London, 1967), discusses the role of vector analysis in Maxwell's *Treatise* and the relationship between vector analysis and the development of electromagnetic theory in the nineteenth century.

Maxwell's methodological statements have received much comment. The classic analysis, strongly antagonistic, is by P. Duhem: *The aim and structure of physical theory*, trans. P. P. Wiener (Princeton, 1954), pp. 55–104. More recent discussions include J. Turner, 'Maxwell on the method of physical analogy', *British Journal for the Philosophy of Science* 6 (1955), 226–38, and 'Maxwell on the logic of dynamical explanation', *Philosophy of Science* 23 (1956), 36–47; R. Kargon, 'Model and analogy in Victorian science: Maxwell and the French physicists', *Journal of the History of Ideas* 30 (1969), 423–36; M. B. Hesse, 'Logic of discovery in Maxwell's electromagnetic theory', in *Foundations of scientific method: the nineteenth century*, ed. R. N. Giere and R. S. Westfall (London, 1973), pp. 86–114; A. F. Chalmers, 'Maxwell's methodology and his application of it to electromagnetism', *Studies in History and Philosophy of Science* 4 (1973), 107–64; and D. M. Siegel, 'Completeness as a goal in Maxwell's electromagnetic theory', *Isis* 66 (1975), 361–8. G. E. Davie, *The democratic intellect: Scotland and her universities in the nineteenth century*, 2d ed. (Edinburgh, 1964), pp.

192–7, suggests the influence of Scottish 'commonsense' philosophy on Maxwell's view of analogy; this interpretation is amplified by R. Olson, *Scottish philosophy and British physics, 1750–1880: a study in the foundations of the Victorian scientific style* (Princeton, 1975), pp. 287–321, a discussion of the role of Scottish philosophy in shaping the use of hypotheses by British physicists.

The development of electrodynamics by Maxwell's British followers has received little discussion. J. Z. Buchwald, 'The Hall effect and Maxwellian electrodynamics in the 1880s', *Centaurus* 23 (1979–80), 51–99, 118–62, reassesses Maxwell's theory of charge and its interpretation by British physicists. H. Stein, ' "Subtler forms of matter" in the period following Maxwell', in *Conceptions of ether*, ed. G. N. Cantor and M. J. S. Hodge (Cambridge, 1981), pp. 309–40, reviews ether theories. D. R. Topper, 'Commitment to mechanism: J. J. Thomson, the early years', *Archive for History of Exact Sciences* 7 (1971), 393–410, discusses Thomson's mechanical view of nature. Joseph Larmor, *Aether and matter: a development of the dynamical relations of the aether to matter on the basis of the atomic constitution of matter* (Cambridge, 1900), gives a contemporary perspective on ether theory at the turn of the century. Larmor's papers on the ether are collected in his *Mathematical and physical papers,* 2 vols. (Cambridge, 1929). Larmor collected Fitzgerald's papers in his edition of *The scientific writings of the late George Francis Fitzgerald* (Dublin, 1902). J. H. Poynting's *Collected scientific papers* (Cambridge, 1920) and Oliver Heaviside's *Electromagnetic theory,* 3 vols. (London, 1893–1912), are major sources.

On German electrodynamics, K. L. Caneva, 'From galvanism to electrodynamics: the transformation of German physics and its social context', *Historical Studies in the Physical Sciences* 9 (1978), 63–159, surveys developments in the first half of the nineteenth century. K. H. Wiederkehr, *Wilhelm Eduard Weber* (Stuttgart, 1967), is a biographical study. Weber's theory is accessible in his 1848 paper, translated in *Scientific memoirs*, vol. 5, ed. R. Taylor (London, 1852; reprint ed., New York, 1966), pp. 489–529. His major papers on electricity are collected in the third volume of his *Werke,* 6 vols. (Berlin, 1892–4). A. E. Woodruff, 'Action at a distance in nineteenth-century electrodynamics', *Isis* 53 (1962), 439–59, discusses Weber. M. N. Wise, 'German concepts of force, energy and the electromagnetic ether: 1845–1880', in *Conceptions of ether*, ed. G. N. Cantor and M. J. S. Hodge (Cambridge, 1981), pp. 269–307, is a valuable analysis of German electrodynamics from Weber to Helmholtz. R. McCormmach, 'Hertz', *Dictionary of scientific biography*, 6:340–50, gives an excellent survey of Hertz's work. T. K. Simpson, 'Maxwell and the direct experimental test of his electromagnetic theory', *Isis* 57 (1966), 411–32, reviews the context of Hertz's experiments. S. D'Agostino, 'Hertz's researches on electromagnetic waves', *Historical Studies in the Physical Sciences* 6 (1975), 261–323, provides a detailed account of Hertz's experiments and their theoretical implications. P. M. Heimann, 'Maxwell, Hertz and the nature of electricity', *Isis* 62 (1971), 149–57, discusses Hertz's interpretation of Maxwell's theory. H. Hertz, *Electric waves: being researches on the propagation of electric action with finite velocity through space*, trans. D. E. Jones (London, 1893; reprint ed., New York, 1962), is a collection of Hertz's major papers on electricity.

T. Hirosige gives an excellent account of aberration and the ether in 'The ether problem, the mechanistic world view and the origins of the theory of relativity', *Historical Studies in the Physical Sciences* 7 (1976), 3–82. D. B. Wilson, 'George Gabriel Stokes on stellar aberration and the luminiferous ether', *British Journal for the History of Science* 6 (1972), 57–72, and L. S. Swenson, *The ethereal ether: a history of the Michelson–Morley–Miller aether-drift experiments, 1880–1930* (London, 1972), provide further discussion of the ether problem.

Lorentz's major papers on electrodynamics are reprinted in vols. 2 and 5 of his *Collected papers*, 9 vols. (The Hague, 1934–9). The development of Lorentz's electrodynamics is analysed in depth by T. Hirosige in 'The origins of Lorentz's theory of electrons and the concept of the electromagnetic field', *Historical Studies in the Physical Sciences* 1 (1969), 151–209. Hirosige's 'Electrodynamics before the theory of relativity, 1890–1905', *Japanese Studies in the History of Science* 5 (1966), 1–49, surveys the work of Lorentz and his contemporaries. R. McCormmach, 'H. A. Lorentz and the electromagnetic view of nature', *Isis* 61 (1970), 459–97, clarifies the conceptual framework of Lorentz's theory and its physical implications. McCormmach's article 'Lorentz' in the *Dictionary of scientific biography*, 8:487–500, gives an excellent account of Lorentz's work.

Some wider dimensions of ether theory are illustrated by D. B. Wilson, 'The thought of late Victorian physicists: Oliver Lodge's ethereal body', *Victorian Studies* 15 (1971), 29–48, and by B. Wynne, 'Physics and psychics: science, symbolic action and social control in late Victorian England', in *Natural Order: historical studies of scientific culture*, ed. B. Barnes and S. Shapin (London, 1979), pp. 167–86.

Matter Theory: Problems of Molecular Physics

A. W. Thackray has collected his papers on Dalton in *John Dalton: critical assessments of his life and science* (Cambridge, Mass., 1972). An alternative perspective is offered by H. Guerlac, 'The background to Dalton's atomic theory', in *John Dalton and the progress of science*, ed. D. S. L. Cardwell (Manchester, 1968), pp. 57–91. A. J. Rocke, 'Atoms and equivalents: the early development of the chemical atomic theory', *Historical Studies in the Physical Sciences* 9 (1978), 225–63, analyses the assimilation and status of the chemical atomic theory and reviews the literature on this subject. W. H. Brock and D. M. Knight, 'The atomic debates', *Isis* 56 (1965), 5–25, and W. H. Brock, ed., *The atomic debates* (Leicester, 1967), focus on the debates of the 1860s on chemical atomism. G. Buchdahl, 'Sources of scepticism in atomic theory', *British Journal for the Philosophy of Science* 10 (1959), 120–34, and M. J. Nye, 'The nineteenth-century atomic debates', *Studies in History and Philosophy of Science* 7 (1976), 245–68, discuss the nineteenth-century scepticism towards the atomic theory. The two volumes of *Classical scientific papers: chemistry*, ed. D. M. Knight (London, 1968, 1970), provide facsimiles of original papers on atomism and the chemical elements. Studies on the problem of structure in organic chemistry include J. H. Brooke, 'Organic synthesis and the unification of chemistry: a reappraisal', *British Journal for the History of Science* 5 (1971), 363–92, and 'Laurent, Gerhardt and the philosophy of chemistry', *Historical Studies in the Physical Sciences* 6 (1975), 405–29; and S. C. Kapoor, 'The origins of Laurent's organic classification', *Isis* 60 (1969), 477–527.

Papers on the kinetic theory of gases by Clausius, Maxwell, and Boltzmann are collected in *Kinetic theory*, ed. S. G. Brush, 2 vols. (Oxford, 1965–6). Brush's *The kind of motion we call heat: a history of the kinetic theory of gases in the nineteenth century*, 2 vols. (Amsterdam, 1976), is a substantial treatise on gas theory and thermodynamics, a collection of his important papers on these topics, and includes a bibliography of works on the kinetic theory of gases published in the nineteenth century. There are several good studies of kinetic theories of gases in the eighteenth and early nineteenth centuries: G. R. Talbot and A. J. Pacey, 'Some early kinetic theories of gases: Herapath and his predecessors', *British Journal for the History of Science* 3 (1966), 133–49; E. Mendoza, 'A critical examination of Herapath's dynamical theory of gases', ibid. 8 (1975), 155–65; and C. A. Truesdell, 'Early kinetic theories of gases', *Archive for History of Exact Sciences* 15 (1975), 1–66.

Clausius's introduction of probabilistic methods into gas theory is discussed in two papers by I. Schneider: 'Clausius' erste Anwendung der Wahrscheinlichkeitsrechnung' and 'Clausius' Beitrag zur Einführung wahrscheinlichkeitstheoretischer Methoden', *Archive for History of Exact Sciences* 14 (1974–5), 143–58 and 237–61. Maxwell's introduction of probabilities is discussed by C. C. Gillispie, 'Intellectual factors in the background of analysis by probabilities', in *Scientific change*, ed. A. C. Crombie (London, 1963), pp. 431–53, and by E. Garber, 'Aspects of the introduction of probability into physics', *Centaurus* 17 (1972), 11–39. The influence of Clausius on Maxwell is analysed by E. Garber, 'Clausius and Maxwell's kinetic theory of gases', *Historical Studies in the Physical Sciences* 2 (1970), 299–319. H. T. Bernstein, 'J. Clerk Maxwell on the history of the kinetic theory of gases', *Isis* 54 (1963), 206–16, makes available an interesting Maxwell manuscript. P. M. Heimann, 'Molecular forces, statistical representation and Maxwell's demon', *Studies in History and Philosophy of Science* 1 (1970), 189–211, discusses Maxwell's introduction of probabilities and analyses the conceptual framework of his statistical physics.

The problems of molecular physics, spectroscopy, and the equipartition theorem are discussed by E. Garber in 'Molecular science in late-nineteenth-century Britain', *Historical Studies in the Physical Sciences* 9 (1978), 265–97; and M. J. Klein, 'The historical origins of the van der Waals equations', *Physica* 73 (1974), 28–47, reviews the impact of the virial theorem on the formulation of molecular models. There is a biography of Rayleigh, including letters, by R. J. Strutt: *Life of John William Strutt, Third Baron Rayleigh*, rev. ed. (Madison, Wis., 1968). W. McGucken, *Nineteenth-century spectroscopy: development of the understanding of spectra, 1802–1897* (London, 1969), provides a broad survey of the problems of spectroscopy. D. M. Siegel, 'Balfour Stewart and Gustav Robert Kirchhoff: two independent approaches to "Kirchhoff's radiation law" ', *Isis* 67 (1976), 565–600, and H. Kangro, *Early history of Planck's radiation law* (London, 1976), discuss problems of spectra and radiation theory. There are several useful articles on spectroscopy and the chemical elements: W. V. Farrar, 'Nineteenth-century speculations on the complexity of the chemical elements', *British Journal for the History of Science* 2 (1965), 297–323; W. H. Brock, 'Lockyer and the chemists: the first dissociation hypothesis', *Ambix* 16 (1969), 81–99; and R. K. DeKosky, 'Spectroscopy and the elements in the late nineteenth century: the work of Sir William Crookes', *British Journal for the History of Science* 6 (1973), 400–23.

The third (1872) edition of Maxwell's *Theory of heat* (London, 1871) has been reprinted (Westport, Conn., 1970). There is a general discussion by E. E. Daub entitled 'Maxwell's demon', *Studies in History and Philosophy of Science* 1 (1970), 213–27. The statistical physics of Maxwell and Boltzmann is analysed in a penetrating essay by M. J. Klein, 'Maxwell, his demon and the second law of thermodynamics', *American Scientist* 58 (1970), 84–97. Klein gives a more detailed analysis of the development of Boltzmann's statistical interpretation of the second law of thermodynamics in his paper 'The development of Boltzmann's statistical ideas', *Acta Physica Austriaca, Supplement 10* (1973), 53–106. Klein's *Paul Ehrenfest: the making of a theoretical physicist* (Amsterdam, 1970) includes valuable discussion of Boltzmann and other nineteenth-century themes. The problems of statistical physics are discussed by S. G. Brush in two important papers: 'Foundations of statistical mechanics, 1845–1915', *Archive for History of Exact Sciences* 4 (1967), 145–83, and 'The development of the kinetic theory of gases: VIII. Randomness and irreversibility', ibid. 12 (1974), 1–88, both reprinted in Brush's *The kind of motion we call heat* (Amsterdam, 1976), 2:335–85 and 543–654. There is further discussion by S. G. Brush in 'Irreversibility and indeterminism: Fourier to Heisenberg', *Journal of the History of Ideas* 37 (1976), 603–30. Whereas Brush explores the connections between molecular physics and thermodynamics, P. Clark, 'Atomism versus thermodynamics', in *Method and appraisal in the physical sciences*, ed. C. Howson (Cambridge, 1976), pp. 41–105, emphasises their conceptual disjunction. There is a classic discussion of Boltzmann's statistical ideas by P. Ehrenfest and T. Ehrenfest, *The conceptual foundations of the statistical approach in mechanics*, trans. M. J. Moravcsik (Ithaca, N.Y., 1959). Boltzmann's arguments are accessible in his *Lectures on gas theory*, trans. S. G. Brush (Berkeley, 1964). Boltzmann's papers are collected in his *Wissenschaftliche Abhandlungen*, ed. F. Hasenöhrl, 3 vols. (Leipzig, 1909). There is a general study by R. Dugas, *La théorie physique au sens de Boltzmann* (Neuchâtel, 1959).

Useful accounts of Gibbs's work are M. J. Klein, 'The early papers of J. Willard Gibbs: a transformation of thermodynamics', in *Human implications of scientific advance*, ed. E. G. Forbes (Edinburgh, 1978), pp. 330–41, and 'Gibbs', *Dictionary of scientific biography* 5:386–93; for further discussion see E. Garber, 'James Clerk Maxwell and thermodynamics', *American Journal of Physics* 37 (1969), 146–55. There is a biography by L. P. Wheeler, *Josiah*

Willard Gibbs (New Haven, 1952). On chemical thermodynamics, F. L. Holmes, 'From elective affinities to chemical equilibria', *Chymia* 8 (1962), 105–45, surveys the background to the emergence of physical chemistry; and R. G. A. Dolby, 'Debates over the theory of solution: a study of dissent in physical chemistry', *Historical Studies in the Physical Sciences* 7 (1976), 297–404, provides a detailed account of some of the controversies engendered by the theory of solutions. The article by E. N. Hiebert and H.-G. Korber entitled 'Ostwald', *Dictionary of scientific biography*, 14:455–69, is a detailed study of its subject, with a full bibliography. E. N. Hiebert, 'The energetics controversy and the new thermodynamics', in *Perspectives in the history of science and technology*, ed. D. H. D. Roller (Norman, Okla., 1971), pp. 67–86, reviews the conflict among Ostwald, Planck, and Boltzmann over the conceptual structure of thermodynamics. M. J. Klein, 'Planck, entropy and quanta, 1901–1906', *Natural Philosopher* 3 (1963), 83–108, gives a full analysis of the development of Planck's views on the second law of thermodynamics.

Epilogue

An essay by M. J. Klein, 'Mechanical explanation at the end of the nineteenth century', *Centaurus* 17 (1972), 58–82, explores the arguments of Hertz and Boltzmann in asserting the programme of mechanical explanation. The arguments of the main protagonists are accessible in translation. H. Hertz, *The principles of mechanics presented in a new form*, trans. D. E. Jones (London, 1899; reprint ed., New York, 1956), is a classic work; and the collection of essays by L. Boltzmann, *Theoretical physics and philosophical problems*, ed. B. McGuinness (Dordrecht, 1974), is a selection of his famous popular articles. E. Mach, *The science of mechanics: a critical and historical account of its development* (La Salle, Ill., 1960), presents the arguments of a leading critic of the mechanical viewpoint. There is a substantial literature on Mach. J. T. Blackmore, *Ernst Mach: his life, work and influence* (London, 1972), gives a general account of Mach's ideas; the symposium on Mach in *Synthese* 18 (1968), 132–301, and the collection *Ernst Mach: physicist and philosopher*, ed. R. S. Cohen and R. J. Seeger, in *Boston Studies in the Philosophy of Science VI* (Dordrecht, 1970), include studies of special topics. The article by E. N. Hiebert, 'Mach', *Dictionary of scientific biography* 8:595–607, includes a full bibliography.

There are a large number of studies of the genesis of relativity theory: Of special relevance is the essay by R. McCormmach, 'Einstein, Lorentz and the electron theory', *Historical Studies in the Physical Sciences* 2 (1970), 41–87, an illuminating analysis of Einstein's physical world view. G. Holton, *Thematic origins of scientific thought: Kepler to Einstein* (Cambridge, Mass., 1973), pp. 165–352, discusses the genesis of the special theory of relativity.

Sources of Quotations

page

1   Stallo, *Concepts and theories of modern physics* (reprint ed., Cambridge, Mass., 1960), p. 60.

15   P. S. de Laplace, *Traité de mécanique céleste,* 5 vols. (Paris, 1799–1825), 5:99.

17   S. D. Poisson, *Mémoires de l'Académie des Sciences* 8 (1829), 361.

24   *Miscellaneous works of the late Thomas Young,* ed. G. Peacock, 3 vols. (London, 1855), 1:383.

26   *Mathematical papers of the late George Green,* ed. N. M. Ferrers (Cambridge, 1871), p. 245; *The collected works of James MacCullagh,* ed. J. H. Jellett and S. Haughton (Dublin, 1880), p. 184.

28   Fourier, *Analytical theory of heat,* trans. A. Freeman (reprint ed., New York, 1955), p. 7.

33   *Faraday's diary, being the various philosophical notes of experimental investigation made by Michael Faraday,* ed. T. Martin, 7 vols. (London, 1932–6), 1:369.

35   Faraday, *Experimental researches in electricity,* 3 vols. (London: 1839–55; reprint ed., New York, 1965), 3:1–2.

36   P. Ewart, *Memoirs of the Manchester Literary and Philosophical Society* 2 (1813), 144, 169; G. Coriolis, *Du calcul de l'effet des machines* (Paris, 1829), p. iv.

40   *The scientific papers of James Prescott Joule,* 2 vols. (London, 1884–7; reprint ed., London, 1963), 1:158, 123.

41   Joule to W. Thomson, October 6, 1848 (Add 7342/J61, Cambridge University Library).

41–3   Helmholtz, *Wissenschaftliche Abhandlungen,* 3 vols. (Leipzig, 1882–95), 1:6, 16, 17, 25.

45   James Thomson, *Collected papers in physics and engineering,* ed. J. Larmor and J. Thomson (Cambridge, 1912), p. xxxi; W. Thomson, *Transactions of the Royal Society of Edinburgh* 21 (1854), 123.

49–51   Thomson, *Mathematical and physical papers,* 6 vols. (Cambridge, 1882–1911), 1:102, 117, 118–19 n.

52–3   Clausius, *Philosophical Magazine* 2 (1851), 4, 104; ibid, 14 (1857), 108.

55   Rankine, *Philosophical Magazine* 2 (1851), 509.

56   Add 7342/PA128, Cambridge University Library, in C. W. Smith, *Archive for History of Exact Sciences* 16 (1976), 280–8; Thomson, *Papers*, 1:175.

57   J. Larmor, *Origins of Clerk Maxwell's electric ideas, as described in familiar letters to William Thomson* (Cambridge, 1937), p. 11.

58–9   Thomson, *Papers*, 1:511; 2:34; Rankine, *Philosophical Magazine* 5 (1853), 106; *Miscellaneous scientific papers of W. J. Macquorn Rankine*, ed. W. J. Millar (London, 1881), pp. 200, 203, 209.

61   M. Faraday, *Experimental researches in chemistry and physics* (London, 1859), pp. 449, 460; Rankine, *Philosophical Magazine* 17 (1859), 250; L. Campbell and W. Garnett, *The Life of James Clerk Maxwell* (London, 1882), p. 289.

63   Helmholtz, *Wissenschaftliche Abhandlungen*, 1:73.

64–5   Rankine, *Papers*, p. 351; Clausius, *Philosophical Magazine* 12 (1856), 96; Clausius, *Annalen der Physik* 125 (1865), 400; ibid, 116 (1862), 79.

66   *The scientific papers of James Clerk Maxwell*, ed. W. D. Niven, 2 vols. (Cambridge, 1890; reprint ed., New York, 1965), 2:664–5.

67–8   Thomson, *Papers*, 1:511; Helmholtz, *Popular lectures* (London, 1889), p. 154; Clausius, *Philosophical Magazine* 35 (1868), 419.

69   Maxwell, *Scientific papers*, 2:376.

70   Thomson and Tait, *Treatise on natural philosophy* (Oxford, 1867), p. vi.

72   Faraday, *Electricity*, 3:30; S. P. Thompson, *The life of William Thomson, Baron Kelvin of Largs*, 2 vols. (London, 1910), 1:215; Larmor, *Origins*, p. 8; Maxwell, *Scientific papers*, 1:527.

73   Maxwell, *Treatise on electricity and magnetism*, 2 vols. (Oxford, 1873), 1:99.

73–8   Faraday, *Electricity*, 1:16, 19, 393, 386, 362; 3:194; 2:290–3; 3:447, 533, 414, 194, 331.

79   Maxwell, *Treatise*, 2:435; *Carl Friedrich Gauss Werke*, 12 vols. (Göttingen, 1863–1933), 5 [1867]:629; Thomson, *Cambridge and Dublin Mathematical Journal* 1 (1846), 92.

82   Thompson, *Life of Thomson*, 1:203, 215; Thomson, *Philosophical Transactions of the Royal Society* 141 (1851), 250.

82–4   Thomson, *Philosophical Magazine* 13 (1857), 198–200; Add 7342/NB35, Cambridge University Library, in O. Knudsen, *Centaurus* 16 (1972), 47–50; Thomson, *Philosophical Magazine* 34 (1867), 15.

85–8   Larmor, *Origins*, p. 17; Maxwell, *Scientific papers*, 1:156–60, 187.

88   Add 7655, Cambridge University Library, in P. M. Heimann, *Archive for History of Exact Sciences* 6 (1970), 181; Campbell and Garnett, *Life of Maxwell*, 2d ed. (1884), p. 203.

89–92 Maxwell, *Scientific papers*, 1:188, 452, 489, 468, 486; C. G. Knott, *Life and scientific work of Peter Guthrie Tait* (Cambridge, 1911), p. 215; Maxwell, *Treatise*, 2:416.

92–4   Maxwell, *Scientific papers*, 1:491, 500, 577; Maxwell, *Treatise*, 2:417; idem, *Scientific papers*, 1:533, 564.

95–8   Maxwell, *Treatise*, 2:184–5, 193–4, 183, 202.

99   *The scientific writings of the late George Francis Fitzgerald*, ed. J. Larmor (Dublin, 1902), pp. 73, 153.

100   Thomson, *Baltimore lectures on molecular dynamics and the wave theory of light* (Cambridge, 1904), pp. vii, 5–13.

101–2 Royal Society, *Referees reports* 12:160 (dated February 5, 1894); Larmor, *Aether and matter* (Cambridge, 1900), pp. vi, 319; idem, *Mathematical and physical papers*, 2 vols. (Cambridge, 1929), 1:517, 631.

105   Maxwell, *Treatise*, 2:438.

108   Hertz, *Electric waves: being researches on the propagation of electric action with finite velocity through space*, trans. D. E. Jones (London, 1893; reprint ed., New York, 1962), p. 107.

109   Ibid., p. 183; Lorentz, *Collected papers*, 9 vols. (The Hague, 1934–9), 9:97.

111   Hertz, *Electric waves*, pp. 21, 28.

112   *Report of the British Association* (1888), p. 561.

129   Clausius, *Annalen der Physik* 105 (1858), 239, 243.

130   Maxwell, *Scientific papers*, 2:343.

131   Ibid., 2:33, 374; Campbell and Garnett, *Life of Maxwell*, pp. 438–9; Add 7655, Cambridge University Library, in P. M. Heimann, *Studies in History and Philosophy of Science* 1 (1970), 201.

133   Maxwell, *Scientific papers*, 2:363.

139–40 Thomson, *Baltimore lectures*, 527; Knott, *Life of Tait*, pp. 214–15; Thomson, *Papers*, 5:12; Maxwell, *Theory of heat*, 3d ed. (London, 1872), pp. 153–4, 308–9.

141   Boltzmann, *Wissenschaftliche Abhandlungen*, ed. F. Hasenöhrl, 3 vols. (Leipzig, 1909), 1:9.

142   Ibid., 2:120.

149   Thomson, *Baltimore lectures*, p. 486.

153–4 Einstein, quoted in R. McCormmach, *Historical Studies in the Physical Sciences* 2 (1970), 56, 60.

Index

affinities (chemical), 14, 18–20
Airy, George Biddell, 25
Alembert, Jean Lerond d', 13
Ampère, André Marie, 31, 34, 44
analogy, physical, 3, 29–30, 79, 85–8, 95
Arago, François, 18
Arrhenius, Svante, 147
atomism: chemical, 19, 120–7; vortex, 83–4, 98, 134
Avogadro, Amedeo, 127

Bernoulli, Daniel, 13, 36, 128
Bernoulli, Johann, 13, 36, 44
Berthollet, Claude Louis, 18, 146
Berzelius, Jöns Jacob, 124–6
Biot, Jean Baptiste, 17, 18, 23
Boltzmann, Ludwig, 8, 138–9, 141–6, 149–52
Boscovich, R. J., 77
Brodie, Benjamin Collins, 126–7
Bunsen, Robert, 135

caloric, 17–18, 44, 48–51
Cannizzaro, Stanislao, 127
Carnot, Lazare, 48
Carnot, Sadi, 4, 45–57; cycle, 48–50
Cauchy, Augustin Louis, 25
Cavendish, Henry, 15, 36
chemistry, 18, 120–5; organic, 125–6; physical, 146–7
Clairaut, A. C., 15
Clapeyron, Emile, 49–51
Clausius, Rudolf, 4, 5, 7, 67, 130; entropy, 65; kinetic theory of gases, 7, 128–9; molecular models, 53, 128,

141; thermodynamics, 4, 5, 52–7, 64–8, 141
Clifford, William Kingdon, 97
Colding, Ludvig, 162
cosmogeny, 66–9; nebular hypothesis, 67
Coulomb, Charles Augustin, 15, 103

Dalton, John, 19, 121–5, 146
Davy, Humphry, 19–20, 33, 37, 124–5
disgregation, 65–6, 141
displacement (of electricity), 92–3, 111
Duhem, Pierre, 169
Dumas, Jean Baptiste, 125–6
dynamical explanation, 26–7, 69–71, 93–9, 101–3, 111, 117, 119, 149, 151
dynamical theory, 9, 56, 94, 130, 131, 139–41
dynamo, 62–3

Einstein, Albert, 10, 153–5
electromagnetism, 3, 6, 30–4, 73, 93, 103–7, 109–12; see also field, concept of
electron, concept of, 102–3, 116–19
energetics, 59, 147
engines (heat), 41, 46–7, 52–4
energy, 1, 4–6, 43, 51, 58–64, 94–5, 98, 101, 143, 147; conservation of, 2, 3, 4, 41–4, 52–64, 66–9; dissipation of, 2, 51–2, 56–8, 66–9, 147
entropy, 5, 8, 64–8, 141–3, 146–8
equipartition theorem, 134–5, 138–9
ether, concept of, 2, 7, 72–3, 78–9, 107, 112; drag, 112–16; elastic solid,

180